T0261349

Innovation Management in the ICT Sector

To Carine and Lauren: the sunshine in my soul.

Innovation Management in the ICT Sector

How Frontrunners Stay Ahead

Edward I. Huizenga

Business architect and management author with Altuïtion, a business innovation consultancy firm, and guest lecturer at Erasmus University Rotterdam, The Netherlands.

Edward Elgar
Cheltenham, UK • Northampton, MA, USA

© Edward Huizenga 2004

All rights reserved. No part of this publication may be reproduced, stored in a retrieval system or transmitted in any form or by any means, electronic, mechanical or photocopying, recording, or otherwise without the prior permission of the publisher.

Published by
Edward Elgar Publishing Limited
Glensanda House
Montpellier Parade
Cheltenham
Glos GL50 1UA
UK

Edward Elgar Publishing, Inc.
136 West Street
Suite 202
Northampton
Massachusetts 01060
USA

A catalogue record for this book
is available from the British Library

Library of Congress Cataloguing in Publication Data

Huizenga, Edward.
    Innovation management in the ICT sector : how frontrunners stay ahead / Edward I. Huizenga.
        p. cm.
    Includes bibliographical references and index.
        1. Information technology--Management. 2. Technoligical innovations--Management. I. Title

    HD30.2.H84 2004
    004'.068'4--dc22

                                                                    2003068787

ISBN 1 84376 567 5 (cased)

Printed and bound in Great Britain by MPG Books Ltd, Bodmin, Cornwall

Contents

Foreword

For a few years now the Information and Communication Technology (ICT) sector has been in the spotlight. We have seen the economic boom of the sector in the 1990s and the recent struggling of many ICT companies to survive. Innovation became and still is a focus in the strategy and survival of many ICT companies, from start-ups to established companies. Are there signs that the sector thrives on innovation and entrepreneurship? In this book an analysis is made of the key success factors affecting innovation in a knowledge and service intensive high-tech sector. The research deals with the strategic and organisational implications of innovative European ICT companies. It does not go into detail on technological innovations. It especially addresses the managerial issues, the challenges and opportunities faced by ICT companies in today's marketplace.

I intended to write this book not as a pure scientific report. From the beginning I assumed that the stories from the European ICT companies were relevant to scientists, managers and advisors as well as for students and teachers. My goal was to explore and explain the state of affairs of the exciting innovation management field. And apply new approaches to the ICT practice to understand what drives these companies to become a frontrunner. Some parts of the book are more interesting to the scientific reader and other parts more attractive for the practitioners. The academic viewpoint of this book includes the development of a pioneering methodology, called the case survey method. A framework, which provides insight into the strategy, the innovation style, the organisation structures and the knowledge development associated with innovation in the ICT sector. As a management writer one is supposed to tell managers how to do the right things (right), to give insights, open up the reader's mind to identify the key to solving problems. This noble intention is tackled in this book by means of thorough research and through business case storytelling. Storytelling is a good ancient 'habit' and is a useful way to illustrate new insights with anecdotes. This book tells the stories and anecdotes of European ICT firms to become and stay a frontrunner in innovation.

Together with my research team we followed the progression of 32 European ICT companies over a period of more than seven years. We were curious to find new insights into innovation of ICT firms in Europe, compared to the stories of Silicon Valley. Let us be clear, the companies themselves have performed the innovative work. We signed up the managerial choices, the developments and compared the companies to analyse innovation patterns and find out how frontrunners perform better. The book tries to provide the followers with suggestions on how to escape from their position and join the frontrunners. The market leaders are provided with insights, which allow them to keep ahead of the pack.

Throughout this research journey I relied on the support of many people. I would like to thank especially the CEOs, marketing and business development managers who have been generous to open their company doors to tell their challenges, their concerns and practices. Special thanks go to my research teammates Jan Cobbenhagen, Wynand Bodewes and Minouk den Hertog. Many scientific colleagues have supported me with the PhD thesis. Jos Lemmink, Corien Gijsbers and of course all the colleagues from MERIT and Altuïtion, thanks for your valuable contributions and the vivid atmosphere. Special thanks go to my friend Friso den Hertog. Thanks for your intellectual challenging support and mentoring. Thanks mate. I want to thank my family. And the two most important people in my life, my two girls Carine and Lauren. You make the sun shine in my soul.

This research programme had many sponsors. The sponsors that co-financed the research programme are the Dutch Organisation for Scientific Research (NWO), MERIT (Maastricht Economic Research Institute on Innovation and Technology), Maastricht University, Ministry of Economic Affairs, the ICT sector organisation FENIT, Altuïtion, and the ICT companies.

Edward Huizenga
's-Hertogenbsoch, The Netherlands, 2004

Author's Bibliography

Dr Edward Istvàn Huizenga is business architect and management author with Altuïtion, a consulting firm specialised in business innovation. He has written several books and articles on (new) business development and innovation and is the co-author of *The Knowledge Enterprise: Implementation of Intelligent Business Strategies*. Edward Huizenga has been involved in major consulting assignments at boardroom level in the area of strategy, marketing and (new) business development. He is known for his innovative ideas and for challenging conventional wisdom. Edward Huizenga received his Master's Degree from Maastricht University, The Netherlands, and Universidad de Zaragoza, Spain, and a PhD on innovation management from Maastricht University. He was a Research Fellow at Maastricht Economic Institute on Innovation and Technology (MERIT), Maastricht University. He is a guest lecturer at Erasmus University Rotterdam and speaker at corporate level. He can be contacted at: edwardhuizenga@hotmail.com.

Other book publications:

2001   *Innovation Management: How Frontrunners Stay Ahead, An Empirical Study on Key Success Factors in the ICT Sector*, E.I. Huizenga, Datawyse - Universitaire Pers, Maastricht.

2000   *The Knowledge Enterprise: Implementation of Intelligent Business Strategies*, F. den Hertog & E. Huizenga, Imperial College Press, London.

1997   *The Knowledge Factor: Competing as a Knowledge Enterprise* (in Dutch), F. den Hertog & E. Huizenga, Kluwer BedrijfsInformatie, Deventer.

1997   *The Innovative Software Enterprise: Strategy, Organisation and Human Resource Management* (in Dutch), F. den Hertog & E. Huizenga, Kluwer BedrijfsInformatie, Deventer.

PART I:

Innovation Management

1. Exploring Innovation in the ICT Sector

1.1 INTRODUCTION

The twenty-first century has begun as a turning decade for managerial and academic thinking on innovation management in high-tech sectors. Innovation is important in today's markets and can dramatically improve a company's competitive position, in particular in the ICT sector. The Internet and electronic commerce represented such significant innovations. New Internet start-ups and established dotcom companies regularly appeared in the news, Internet service providers and online retailers and financial service providers started initial public offerings in Europe, the United States, and the NASDAQ has been and still is the most volatile high-technology index on the Stock Exchange. Are these clear signs that the sector seems to thrive on innovation and entrepreneurship?

In the management literature it is observed that innovation drives corporate success and is a strategic endeavour contributing to a firm's differentiating capacity. Some researchers even say that innovation has become the primary process (Jaikumar, 1986). Recently studies (e.g. Huizenga, 2001; Cooper & Kleinschmidt, 1996; Brown & Eisenhardt, 1995; Montoya-Weiss & Calantone, 1994) have investigated the key success factors and sources of attaining competitive advantage through innovation. Such research has started to contribute to new insights into how a company turns good ideas into money and masters the art of designing and developing new products, services and processes. Meanwhile, companies need to ensure that resources are leveraged for the creation of new markets, products and services to serve customer needs (e.g. Hamel & Prahalad, 1994).

On the other hand, it has been observed in the management literature that innovation can also be a high-risk adventure. Innovation projects may face a delay due to technical complexity, original ideas fail to make it to the market, or customers do not appreciate new products or services. These risks are well documented in the empirical and theoretical literature (e.g. Stacey, 1996; Rothwell, 1994; Roussel *et al.*, 1991; Imai, Nonaka &

Takeuchi, 1996), as well as the opportunity risks associated with refraining from new product and process development. Firms that ignore renewal might reinforce the existing practices and hamper innovation (e.g. Dougherty, 1996). The consequence will eventually be a loss of revenue and market position. The danger of neglecting innovation can have a serious impact on the long-term survival of companies. Empirical findings have revealed that the inability to connect innovation with organisational resources, processes, and strategy thwarts innovation outcomes (e.g. Eisenhardt & Tabrizi, 1995; Dougherty, 1990; Burgelman, 1983).

It is increasingly recognised by the business and academic communities alike that the nature of innovation and the process of developing new products and services increasingly involve a mixture of managing potential 'good fortune', complexity and high risk. Furthermore it is recognised that service dimensions enter into all products and businesses. In fact, 'service functions' seem to play a role in all business firms (e.g. Coombs *et al.*, 1999) and services businesses are not product business (Nambisan, 2001). This raises new challenges for managers, as their competitive environment becomes more complex as a result of the rise of the 'service intensity'. More recently, a view has begun to emerge investigating innovation in dynamic high-tech, high-risk and knowledge-intensive sectors, e.g. the information and communication technology sector (ICT). Before discussing this subject let us briefly address some of the typical characteristics associated with innovation.

Innovation characteristics
To be effective in their innovation efforts, companies have to take into account the nature of the strategy and organisational processes of innovation. Innovation is not similar to performing regular 'business as usual'. Innovation is to a large extent unpredictable, complex, dynamic, non-routine-based, and involves creativity and risk. Innovation is difficult to control and is essentially an entrepreneurial process as such. Among the basic characteristics of innovation we identify (based on Tushman & Moore, 1992; Van der Ven, 1986) the following:

Managing the chance, the risks and the change
These elements accompany the outcome of innovation, whether it be a new product, service or new business process. Innovation is a risky adventure involving a great deal of unforeseeable uncertainties. Innovation is based on a promising idea that is partially associated with luck and chance as to whether the outcome will be accepted by customers. Development is

associated with high-risk trajectories with potential pitfalls and uncertain, high cost and results.

The complex balance of customer interaction

Innovation processes are difficult to manage because of the range of complexities involved, such as functional interfaces, overlapping activities and customer input and involvement. Innovation requires multiple people, disciplines, and activities that need to be integrated. Innovation activities are a mixture of creativity, irrationality and feasibility. They are typically unstructured activities, which need to be aligned with a commercial rationality. Quinn (1985) pointed out that innovation management resembles the management of a creative chaos. Innovation balances between tight managerial co-ordination and commitment to free-floating experimentation to discover new technological or market opportunities. The multidisciplinary nature places strong demands on the cost, benefit and timing of involving customers in innovation.

New business development: the paradoxes and dynamics

Van der Ven (1986) stated that innovation is conducted in an enterprise where operational business processes ('steady-state processes') appear to be in stability and formal control. Yet at the same time enterprises are facing a paradox because new alternatives beyond the steady state need to be developed (e.g. Abernathy & Clark, 1985): new business development. This requires them to move away from formal control and stability. Innovation absorbs the resources that have to be distracted from 'doing business as usual' that currently render value. These resources have to be devoted to unknown new business opportunities that might render value in the long run. This duality of innovation might put pressure on the willingness to innovate and introduces control problems.

The cultural resistance to renewal

Innovation is a 'people business'. If the people in the organisation are motivated to change, this will enhance the innovation success of the firm. Resistance to change, however, can hamper the outcome. It drives out the existing practices, structures and routines in the organisation. Work routines have to be abolished and unlearned and new routines have to be learned. Innovation management is therefore also about developing new routines and meanwhile preventing core rigidities from occurring (e.g. Leonard-Barton, 1992).

1.2   RESEARCH GOALS & STRUCTURE OF THE BOOK

In this book we will tell about the issues arising from the state of affairs on innovation in the European ICT sector and also tell what separates winners from losers. For this purpose, we chose a sector characterised by rapid technological developments and changing market demands. Moreover, the ICT sector is a knowledge-intensive sector where the service intensity might play an increasingly important role. A company that operates under these circumstances is confronted with fierce competition, short product life cycles and the pressure to differentiate, creating a sense of urgency to innovate.

It is expected that zooming into innovation in such a sector will increase our knowledge of the effectiveness of innovation management. It will provide insight into the company level drivers for successful innovation. Exploring the role of strategy and organisation can contribute to further theory building on innovation management. The research builds on the following main research questions:
- Which key success factors contribute to a high innovation performance in the ICT sector?
- Why and how do frontrunners differentiate from members of the pack?
- To provide building stones for management practice and theory, what explanations from strategy and organisation theory can support our understanding of the conditions for successful new product and service development?

The ICT innovation study is founded on two pillars, a management viewpoint and a scientific viewpoint. The management viewpoint is intended to provide firms with clear insights, arguments and tools to improve or sustain their innovation success. This research has led to various management publications (e.g. Huizenga, 2001; Den Hertog & Huizenga, 2000, Cobbenhagen, 1999; Den Hertog & Huizenga, 1997; Den Hertog & Huizenga, 1997; Cobbenhagen, Den Hertog & Pennings, 1995).

The cycling metaphor
It is important to note that we used a cycling metaphor in this book to describe the differences in performance between firms. In cycling terms we designate firms as frontrunners, members of the pack or laggards. The viewpoint focuses on the practices of frontrunners, how they perform, how

they differ and how they sustain in their innovation success. These subjects are addressed by means of the key success factor framework.

It is important to note that the study is not focused on statistically measuring the difference between the best performing firms (the frontrunners) and the worst performing firms (pack members and laggards), but on identifying and explaining what separates the winners. We try to provide suggestions on how to escape from a pack position and join the frontrunners. The frontrunners are provided with insights, which allow them to keep ahead of the pack. We will attempt to show, step by step, how successful organisations approach innovation and achieve (winning) results.

This book dwells upon a large-scale research study among European ICT firms. The results of this study are sometimes provocative, bringing new perspectives on innovation in a knowledge and service intensive sector. Sometimes the results confirm existing knowledge and insights on innovation. Before we get into detail of the worthwhile innovation theories and the empirical findings we present some background to the study.

Chapter 1 describes the state of affairs of the European ICT sector during the period of the research of 1995 to 2001. In chapter 2 we discuss the most worthwhile insights from current research on innovation. Special attention is paid to valuable innovation frameworks and contributions to understanding the success factors for innovation. Chapter 3 explores the strategic nature of firms and the implications for innovation. The most promising strategy perspectives, which incorporate the role of innovation are debated. Chapter 4 discusses the organisation and process of innovation. Chapter 5 is an intermediate chapter explaining the research methods. Chapter 6 through 9 brings the theories to the actual management practices of innovation. Success factors are explored and discussed with business case material.

1.3 THE CONTEXT OF THE RESEARCH STUDY: THE INFORMATION AND COMMUNICATION TECHNOLOGY SECTOR

In 1996, in the beginning of the research period, the European Information Technology Observatory (EITO, 1996) reported the following on the importance of new product, service and process development:

The outlook for 1996 and following years is positive, but of course this does not mean that conditions have become easier for information and communication technology companies in Europe. Heightened global competition puts pressure on all companies, technological evolution forces continuous innovation in products and organisations, so that restructuring is still necessary to adjust the balance between revenue and costs. The 1995 results provide a clear demonstration of the ability of companies to adapt successfully the products and services they offer, to enter into new business, to establish new alliances and to learn new skills. Similarly, and even to a greater extent, flexibility and the ability to be innovative will be needed for the coming years'. And later: 'Developing entrepreneurial ideas and innovative applications is the challenge for ICT companies. As the market growths and technological achievements create possibilities for new markets, products, services and distribution channels, entrants will turn their attention to the ICT market. Whereas the sector has for many years been dominated by technological development, currently development processes and the success of products and services is market driven. Customers, end users and suppliers are of great importance to the development of the sector. This indicates that operating boundaries are shifting and new competitive structures arise where new alliances and partnerships are formed to secure the continuous delivery of innovative applications that are valued by the market.

In 2000 the EITO observed that the sector emerges in a new wave of European entrepreneurialism in the high-tech industry, particularly since 1998 with the emergence of players like Amazon.com and the many start-ups in E-commerce and the massive demand for web-enabling capabilities. Furthermore, a main breakthrough is expected from the development and revolution in the mobile communications driven by the diffusion Internet-enabled devices, such as WAP mobile phones and palm-tops. One might expect innovation to be of strategic importance to the information and communication technology (ICT) sector.

The service sectors are among the major users of new ICT. According to Coombs *et al.* (1999) about three quarters of all expenditure on ICT hardware and software in the UK and USA stems from services. Certain types of retail and financial services are the leading users, both in terms of the volumes of hardware and software used, and in terms of pioneering new applications and advanced equipment. Some commentators suggest that ICT represents a innovation revolution for all kinds of intermediary services sectors, e.g. brokers in banking, insurance, travel and retail sectors.

In this context it is important to learn more about the innovation efforts performed by ICT companies. However, there is a blank spot in the research field of innovation management with regard to the ICT sector. Only few studies (e.g. Eisenhardt & Tabrizi, 1995; Cusumano & Selby 1995; Iansiti & Clark, 1995; Bruce *et al.*, 1995) have investigated the innovation performance of ICT companies. Little is recorded about what is conducive to the success of innovation at the firm level.

1.4 ICT INDUSTRY AND ICT SERVICE SECTOR: FIGURES & TRENDS

To provide a clear picture of the size and structure of the ICT industry during our research period, we present some financial figures of the sector. The market value of the total information and communications technology (ICT) market will reach 456 billion EURO in Western Europe in 2000 (see table 1.1), a growth of 7.5% compared to 1999 figures (424 billion EURO). Based on the most recent figures this represents approximately 5% of total European Gross Domestic Product (GDP) (EITO, 1999). Of the total ICT expenditure, the IT component will account for 231 billion EURO. Computer hardware accounts for 97 billion EURO (21% of the total ICT value), services and software for 87 and 48 billion EURO respectively, representing 19% and 11% of the total ICT value in Western Europe. The remaining part of total ICT spending can be assigned to telecommunications equipment and services with a market value of 224 billion EURO.

Table 1.1: ICT market size at research period (Source: EITO, 1999-2001)

	Value 2000 Europe (in millions of €)	As % of ICT
Total IT	**231,967**	**51**
IT hardware	96,809	21
IT Services	86,695	19
IT Software	48,463	11
Total Telecom (1)	**224,151**	**49**
TOTAL ICT (2)	**456,118**	**100**

Notes:
(1) both telecom equipment (networks, terminals, switching etc.) and telecom services;
(2) total ICT value is the sum of the total IT and Telecom value.

Growth trends

The European growth figures for each market/product segment showed that software is a driving force for IT growth. For the successive year until 2000 the software growth rate was around 10 to 12%, compared to a services growth of 12% and hardware growth of 5 to 6%.

There are clear signs that growth rates in the hardware sector are influenced by the increased demand for client-server solutions. Also, the increased quality, speed, reliability and performance of computers boosts hardware sales. Firms invest in systems and networks and are more confident in relying on ICT for their business processes. The services sector profits from the increased demand for consulting, facilities management, systems and network implementation and operation services. Firms invest more in these areas of ICT in their search for more effective business processes to beat competitors. The implementation of large complex Enterprise Resource Planning (ERP) and Customer Relationship Management (CRM) software applications has positively influenced IT services spending as well.

Two other business topics related to these service growth figures. First of all, the introduction of the EURO with its consequences for the payments and transactions traffic and services. The convergence of the Western European economies into one monetary union encompassed the introduction of one single currency, the EURO, as the accounting and paying device. This required the adaptation of a large number of computer systems. Secondly, the historical ICT phenomenon of the year 2000, which lead to a growth in the demand for millennium repair services.

The high software growth figures result from many technological and market developments. Among the technological drives are the increased reliability of both software and systems and the introduction of innovative technologies such as Java and HTML. The implementation of client server architectures also positively influences software spending. Another technological trend is related to the fact that companies are entering the stage of integration and standardisation of systems and networks (Nolan, 1973). Firms can invest in tools and applications without having to deal with costly conversions and migrations between systems. One dominant market trend is the demand for relational database software and back-office and front office applications, e.g. Enterprise Resource Planning (ERP) and Customer Relationship Management (CRM) software.

ICT market dynamics

What do these figures mean in terms of changes for the ICT companies? Several structural changes have taken place in the past few years, resulting in new competitive challenges. New competitors have emerged at an accelerating pace, trying to capture some of the market growth. Many .com ventures have emerged that provided a new impetus for the concept of services. New technological ventures were among the potential market entrants in Europe. In today's capital markets informal investors and venture capitalists have found their way to promising start-ups. At the same time established ICT companies have tried to expand their business by aggressively advertising and marketing their products and services. More recently acquisitions and joint ventures have restructured the industry, e.g. the take-overs of small service providers by dominant market players and the take-over of Internet start-ups by large financial and retail banks and insurance companies.

The labour market

The market situation has been influenced also by the state of the labour market in recent years. The demand for ICT personnel, due to high software and ICT services demand, is much bigger than the supply of graduates from universities and (poly)technic schools. This situation has led to the hiring and training of people without an ICT-specific educational background. These market changes have been structural since 1996 and have stimulated ICT suppliers to look for creative solutions such as training unemployed people, hiring people from other ICT-intensive countries such as India, or providing extra bonuses and fringe benefits for potential employees. High-quality skills are the basic resources for the further growth of the ICT sector. Investments in education and training represents one of the strategic areas of investment for the industry and government, especially because of the massive use of ICT, multimedia and the Internet.

These observations give us reason to believe that the rapid changes in products, services and process in the ICT sector are important to consider and represent an interesting research area with respect to innovation. Given the unarguable growth in the importance of ICT products and service sectors, increasing numbers of researchers and managers have to take a fresh look at innovation. This includes questioning the available wisdom regarding the innovation capacity of these firms. But the ICT market seemed unstable at that time due to the rapid technological changes, the rapid emergence of new entrants, and the scarcity of labour skills. Would a phase of a shake-out emerge and would it impact on the innovation drive of firms?

2. Key Success Factors in Innovation Management: Recent Contributions

Effective innovation management has recently gained a lot of research interest (e.g. Harvard Business Review, special issue on innovation, 2002; Cooper & Kleinschmidt, 1996, 1995; Griffin & Page 1996; Souder, 1987; Rothwell, 1972), in particular in studies exploring the factors that distinguish between success and failure. Early economic-oriented insights argued that the size of R&D expenditure might explain the performance differences between firms (see Freeman, 1988). However, more recent research has indicated that there might be other intervening factors that are of even greater importance to innovation performance. Such factors might reside in the way processes are designed, activities are organised and conducted, resources are allocated and strategic objectives are pursued. Further research has investigated project failures and factors that inhibit innovation in order to understand what drives a good idea into a bad product outcome (Abernathy & Clark, 1985).

2.1 INNOVATION: CONCEPTS AND DEFINITIONS

To structure the research domain of innovation, we can split it along two research streams. The difference between these streams is similar to what Mohr (1982) has identified as *process-oriented* and *variance-oriented research*. *Process-oriented* research primarily concentrates on the analysis of the structure of innovation processes. The goal of process-oriented research is to develop a process theory that explains the effectiveness of innovation from the institutional context and the sequence of events in a development process (Den Hertog & Van Sluijs, 1995). The research domain of innovation has generated standard classifications of innovation processes resulting from analysing and comparing the stream and coherence of development activities conducted. The studies belonging to the process-oriented stream have addressed the issue of how innovation is performed. To do so, a process perspective on innovation is needed to analyse 'the temporal sequence of events' and describes the dynamics of change (Van de Ven & Huber, 1990). A number of authors (Pettigrew, 1990; Mohr, 1982) refer to this as the 'process theory' as it explores the context, content and process of change in a time perspective. *Variance-oriented* research mainly concentrates on identifying the success factors in innovation and on analysing what causes the difference between successful and unsuccessful innovations. Variance studies rather focus on the identification of key success factors to explain differences in innovation performance at the product, project and company levels (Van de Ven & Poole, 1990).

Classification schemes of innovation
The discipline of innovation management has rapidly developed into a separate research field throughout the past years. It can be regarded as one of the most rapidly growing fields in terms of the attention devoted by both economics, strategic management and organisational sciences to the role of innovation. This has generated a huge volume of theoretical and empirical work. The literature continues to grow, adding a multidisciplinary character to the innovation management research. The source of this interest can be found in the increasing belief that firms can proactively adapt to changes in their environment by renewing their products, services, processes and organisation. This multidisciplinary interest has the positive effect of generating a rich amount of knowledge on what product and process development is about and why some firms perform better than others. On the other hand, the research field faces the consequence of numerous variations in concepts and topics. Several authors have argued (Brown & Eisenhardt, 1995; Montoya-Weiss & Calantone, 1994; Cooper & Kleinschmidt, 1993; Downs & Mohr, 1976) that the innovation field is characterised by a multiformity of definitions, of designs and units of analysis. Few findings have been accumulated despite the many studies, which has resulted in instability in the research field. This has led to the development of a large number of sub-theories rather than the integration of empirical findings (Damanpour, 1991). This multiformity and sub-theoretical focus is immediately reflected by the numerous innovation definitions. Daft (1983) distinguishes between administrative and technical innovation. Zmud (1982) separates initiation from implementation of innovation to emphasise the adoption of a new product idea. Gobeli and Brown (1988) attempted to structure the field, in which they basically identified four types of classification schemes of innovation.

First, what is the initial *focus* of an innovation? Biemans (1992) refers to innovation as the process of developing a new item and the process of adopting the item and the new item itself. Gobeli and Brown (1988) and Abernathy and Utterback (1978) make a distinction between product, service and process innovation. A distinction, which is assumed to be relevant to the ICT sector. Another distinction is between organisational innovation and technological innovation. Organisational innovation involves the development and transformation of organisational structures and processes. Technological innovation, as defined by Freeman (1988), is the process of technical, design, manufacturing, management, and commercial activities involved in the marketing of a new or improved product or service. Whereas the former is expected to be relevant to IT service providers, the latter is expected to be relevant to software firms. Second, what is the *state of the innovation system*? Innovation can be a programmed range of activities or it can be non-programmed as a process of 'creative destruction' of existing structures and processes. Firms may want to conduct non-programmed innovation in the hope to find new breakthroughs that originate from creativity or slack resources. Third, what is the *source* of innovation? One can distinguish, for example, market pull versus technology push innovations. Both are assumed to be of great importance to the ICT sector. Fourth, what *impact* does an innovation have? This refers to the impact on existing structures and processes. For a firm, an innovation can be *incremental*, implying small improvements, or *radical*, which means it has a radical impact on existing structures. The degree of change and the benefit of the innovation determine the intensity of change (Gobeli & Brown, 1988). Table 2.1 presents a matrix that distinguishes four levels of effects, i.e. application, radical incremental and technological change.

Table 2.1: The effect of an innovation (Source: Gobeli & Brown, 1988)

	low benefits	high benefits
high level of change	application	radical
low level of change	incremental	technological

This matrix is useful in clarifying the complexity associated with ICT innovation. Schumpeter (1942, p. 84) states that radical innovations:

> '... command a decisive cost or quality advantage that strikes not at the margin of the profits and the outputs of existing firms, but at their foundations and their very lives ...'.

Tushman and Anderson (1986) refer to radical as the technological discontinuities that disrupt existing structures and new products differ in every aspect from the existing technological base. Both statements clarify the radical impact attributed to innovation and the consequences for 'doing business as usual'. New insights into the role of competencies and the impact of innovation on competencies have emerged. Researchers argue that innovations destroy competencies and favour newly developed technological competencies at the expense of market incumbents who exploit established competencies (Leonard-Barton, 1995; Tushman & Anderson, 1986).

Abernathy & Clark (1985) describe the strategic impact of innovation in terms of disrupting or conserving competencies and distinguish four types of innovation. New products can require a reorientation of corporate goals or production facilities. These *architectural innovations* depart from the existing technological endowment of the firm and induce the need for a business (re)focus. They are radical in nature because of their durable and dominant designs. They have an overall impact on an industry's structure. Architectural innovations often render existing capabilities obsolete and require new routines to integrate and co-ordinate business activities. They provide new business opportunities by creating new markets. *Market niche innovations* differ from architectural innovation in terms of their basis on existing technology to create new market linkages. *Regular innovation* involves change that affects cost and performance structures. It has no dramatic impact upon existing structures and production systems. These minor innovations are executed within existing markets and competencies. They support and enhance the existing knowledge base of the firm. Typical innovations include changes in product characteristics and quality improvements of services. *Revolutionary innovations* apply to current markets but disrupt the existing technological competence base. These innovations replace existing dominant designs in an industry by designs that are based on novel technologies.

2.2 KEY SUCCESS FACTOR STUDIES

The quest for the drivers of innovation success has been a popular research topic for the last decades. The typical studies compare and contrast the characteristics of winning versus losing ideas, projects and business. In doing so they uncover those factors, which discriminate between the two.

Typical areas of KSF research include:
The industry or service sector. Each sector has a set of key success factors that are determined by the characteristics of the sector itself, e.g. chemicals, ICT, banking or retail. Each company in the industry must pay attention to these factors.
Competitive strategy and industry position. Each company's situation within an industry or service sector is determined by its history and current competitive strategy. The company's position in the industry dictates some KSFs.
Environmental factors. Environmental factors are those areas over which an organisation has little control. The organisation must accomplish its mission while riding the tides of environmental change and selection.
Temporal factors. A number of areas of activity become critical for a particular period of time for a company or a sector. Either because something out of the ordinary has taken place, or because there is a temporary unique resource (such as knowledge, insight, routine, etc.) or a resource that is hard to imitate. Over time these factors disappear as they have become common practice in a sector.
Functional management factors. Each management area has success factors associated with functional disciplines, e.g. marketing.

KSF innovation studies, programmes and classification schemes
Although the above classifications are widely used, the innovation management discipline still disputes on which factors matter and influence success. Going back to the beginning of variance-oriented research we identify the roots of key success factor studies in three major empirical studies: the SAPPHO studies, The Stanford Innovation Project (SINPRO), and the New Product Program (NEWPROD). The MERIT programme (Den Hertog *et al.*, 2000; Cobbenhagen, 1999; Den Hertog & Huizenga, 1997, 1997, 2000) shows similarities with the range of NewProd studies.

The SAPPHO project (Rothwell, 1972; Rothwell *et al.*, 1974) analysed the characteristics of successful innovators in technologically advanced companies. The study compared 43 pairs of product success and failures in

the chemical and instruments sector. Among the key results of this study
are:

- understanding of user needs;
- efficient development;
- market attention;
- senior leadership.

The SINPRO studies (Maidique & Zirger, 1985) analysed 70 pairs of
product success and failures, from 21 case studies, in a longitudinal study in
the US electronics. This study also addressed the importance of product
advantage, market attractiveness and organisation. The product advantage
factor referred to the low cost, high quality or uniqueness of the product.
Market attractiveness referred to the presence of a large volume and high-
growth markets. The organisation factors referred to high internal
communication in cross-functional teams and supportive senior
management.

The NewProd programme (Cooper, 1979) sharpened the emerging research
emphasis on product advantage, market attractiveness and internal
organisation. NewProd studies started with comparing new product success
and failure. The programme produced research results on success factors
through replicating the study in many sectors and countries (Cooper &
Kleinschmidt, 1995). These studies are often regarded as the first projects
that showed consistency in results, quality of the methodological framework
and a consistent chain of reasoning (Barclay, 1992). The key contribution of
these studies is its identification of success factors at the project level and
process-related factors (Cooper & Kleinschmidt, 1995). The success factor
investigations have delivered insights from success stories and provided
lessons on what should be stimulated to improve output. The research is
valuable as it offers ideas for improving the practices of innovation and can
guide management in making development efforts successful.

A more recent NewProd study of particular importance was the study by
Cooper & Kleinschmidt (1987). The study analysed 203 products, of which
123 were a success. A total of 125 firms were involved in this study. The
findings revealed that the most important success factor was *product
advantage*. Product advantage offered the firm a superior position relative
to competitors. Product advantage involved a unique superior attribute in
the eyes of the customer. These benefits were not found in competitive
products. The advantage stemmed also from the high performance-to-cost
ratio and the economic advantages to the users. The use of advanced
technology in the design and the low purchasing costs for the customer

explain the advantage. This product advantage rendered value to customers through high quality and innovative product features that helped customers solve problems and contribute to a higher margin for the firm. In addition, the *organisation* of the internal development process was critical to success.

The organisation involved marketing factors, management factors and synergy factors. Essential was the pre-development planning. This up-front homework would include a basic understanding of the user needs, market knowledge and marketing proficiency. This market focus would improve the essential market information gathering. Along the innovation chain this would result in a clear target market, innovative product features. It would also support the vision and consensus on the product concept and the achievement of a strong market launch.

The management factors involved top management commitment to the project, especially from senior management. Also an efficient and well-planned R&D process and good internal communication between groups are key factors. The synergy between the marketing and technology discipline was conducive to success. This involved a good fit between production, research, sales and marketing requirements and the available resources and skills. Although less important, the market conditions were relevant success factors. Markets characterised by growth and relatively weak competition were conducive to new product success. The authors also indicated that markets characterised by dynamic needs, for example a high degree of change in customer needs and wants for a product category, might increase product success through raising opportunities.

Based on the NewProd experiences, Cooper & Kleinschmidt (1995, 1996) developed a conceptual framework to identify the major drivers of new product performance. Based on studies by Shrivastava & Souder (1987), Griffin & Hauser (1996), Johne & Snelson (1988), Song & Parry (1996, 1999) this framework incorporates five classes of critical success factors:

1. *Process*: the firm's new product development process and the specific activities within this process,
2. *Strategy*: the new product strategy as part of the business strategy,
3. *Organisation*: the way projects are organised,
4. *Culture*: the firm's internal culture and climate for innovation,
5. *Commitment*: management involvement and corporate commitment to new product development.

Complementary to the work by Cooper *et al.*, Montoya-Weiss & Calantone (1994) performed a review study and meta-analysis of 47 studies on new product performance. This study, which is the most complete study in the sense that it integrates the major empirical research findings), identifies four major sets of key success factors that drive new product success (table 2.2).

When we go more into detail in the development process itself we observe that researchers point to the importance of sophisticated management methods in order to handle the complexity (Clark & Wheelwright, 1993; Stacey, 1996; Gobeli & Brown, 1988).

Table 2.2: Key Success factors (Source: Cooper & Kleinschmidt, 1995)

---

1. *Strategic factors*
       Product advantage
       Technological synergy
       Marketing synergy
       Resource capacity
       Product strategy
2. *Development process factors*
       Protocol
       Quality of technical and marketing competencies
       Quality of pre-development activities
       Top management support & skill
       Financial and business analysis
       Speed to market
       Costs
3. *Organisational factors*
       Internal and external relations
       Teamwork
       Communication
4. *Market environment factors*
       Market potential
       Environment
       Market competitiveness

Pre-development work

Pre-development work has been mentioned as one key success factor in innovation (Cooper & Kleinschmidt, 1993, 1994, 1996; Rothwell, 1992; Montoya-Weiss & Calantone, 1994). A strong attention for up-front activities involving a strong market orientation, a thorough understanding of the marketing tasks ensures the quality of pre-development. Project management should start with high-quality pre-development work. The proficient execution of these tasks prior to the start of a development stage can result in a better and faster development process. In line with this finding Johne and Snelson (1988) argued that the existence of a formal procedure for a new product development process is related to success. Such procedural items would set clear target market and product requirements prior to development stage.

Concrete planning and go/no go decision points

One might expect proper planning and explicit project plans to reduce the total development time. Roussel *et al.* (1991) argued that planning and scheduling could reduce the risks between the hand-over of the design and development stage. Cooper's (1994) stage gate development process introduces the presence of go/no go decision points in order to monitor the process. Eisenhardt & Tabrizi (1995) found support that the frequent use of milestones accelerates product development and observed that milestones between stages ensure that the project members keep track of their pre-defined budget and time objectives. In line with Eisenhardt & Tabrizi, Gersick (1994) argues that in conditions of rapid technological change milestones are an effective tool for advancement. Additionally, Ancona & Caldwell (1992) found concrete project plans to be related to the motivation of team members (also Gersick, 1994). Their study indicated that deadlines and milestones create a sense of urgency and commitment with the team to achieve pre-set goals. The presence of dedicated project leaders and team members can drive the profitability of the business development efforts.

Project management and project tools

Cooper & Kleinschmidt (1995, 1996) showed that successful firms are characterised by strong project management. The quality of project management included an emphasis on up-front homework, a sharply defined early product definition before development, tough milestones and flexibility to combine stages and decision points when needed. The role of pre-development work includes a detailed market and technical assessment. By assessing the market potential, the customer needs, capital requirement, manufacturability, skills and technical requirements along the process, firms can prevent the failure of starting product development due to inadequate understanding of the innovation chain.

Strong project management is reflected in formal procedures that are made explicit in project management techniques such as project evaluation and review technique (PERT) or critical path method (CPM). These formal planning and control methods are used to ensure alignment of the complicated streams of activities. Clearly defined tasks, responsibilities, skills, and formal authority characterise project management procedures.

Organisation of innovation

Elements like the multidisciplinary teams (e.g. Gobelli & Brown, 1988), the process of problem solving (e.g. Nonaka & Takeuchi, 1996) and the cross-functional interfaces between (e.g. Galbraith, 1994) are considered beneficial to innovation. In a meta-analytic study on organisational innovation, Damanpour (1991) listed several organisational dimensions that moderate innovation performance (see box 2.1). All these factors are considered to improve the innovation performance and shorten the time-to-market of the process.

Box 2.1: Organisational dimensions moderating innovation
Source: Damanpour, 1991

- Specialisation: reflects a greater variety of specialists, provides a larger knowledge base and increases cross-fertilisation of ideas.
- Professionalism: increases boundary-spanning activity, and commitment to adjust the status quo.
- Formalisation: can hamper the flexibility required to facilitate innovation. The lower the degree of formalisation the more open and receptive an organisation is to change.
- Decentralisation: points to the concentration of authority and control to the lowest possible operational level encouraging awareness, commitment and involvement to innovation.
- Slack resources: allow firms to discover and explore new opportunities and detect emerging market needs.
- Communication: both intensive internal and external communication facilitates the scanning of ideas and the rapid dispersion of information. This can result in a more effective use of information and acceleration of process steps.

In particular, some studies (e.g. Cooper, & Kleinschmidt, 1995; Song & Parry, 1996, 1999; Eisenhardt & Tabrizi, 1995; Cobbenhagen, 1999) argue that whenever more responsibilities are delegated towards the project team, the more autonomous a team works. This is in line with Clark & Wheelwright's (1993) argument of a heavyweight project leader. They propose that when more responsibilities are decentralised towards the project teams, teams are more empowered to execute successive innovation steps. The project responsibilities range from initiating the kick-off of a project to managing the team through the concept, design and development stages. Based on a clear set of project responsibilities project leaders would use their authority and control over resource allocations.

Culture
The cultural factor addresses the atmosphere that resides with an organisation. A lot of studies (Scott-Morgan, 1994; Senge, 1990; Souder, 1987) on culture pointed to success factors, e.g. practices that stimulate brainstorm work, idea generation, or R&D scouting time to discover new opportunities. Culture elements that stimulate informal communication and co-operation and reward learning behaviour allow for entrepreneurship. Moss Kanter (1983) and Pfeffer (1994) argue that innovation is stimulated in an environment that has an external open view and tolerates learning from mistakes and comprises little hierarchical levels.

Management commitment
The key success factor literature (e.g. Cooper & Kleinschmidt, 1987; Johne & Snelson, 1988; Griffin & Hauser, 1996) has expressed that management involvement is an essential key attribute for product success. The presence of managerial elements like motivation, dedication and support are associated with high innovation performance. Among the typical elements are management styles that express risk-taking behaviour, a clear dedication to innovation and the reflection of both in the allocation of resources for product development (e.g. Barney, 1991). This argument implies that management commitment is also reflected in the dedication to the role of human resources (Huselid *et al.*, 1997). Nonaka & Takeuchi (1996) and Brown & Eisenhardt (1995) argued for close involvement of senior management in project decisions and in the case of project difficulties, eventually contributing to higher performance.

The above-discussed studies have been summarised in table 2.5 (see at the end of chapter), which shows a clear variation in industries. Some studies have incorporated multiple industries as research focus. Some of them have been replicated in other industry settings to uncover and validate industry similarities and differences.

2.3   KEY SUCCESS FACTORS IN THE INFORMATION & COMMUNICATION TECHNOLOGY SECTOR

Little evidence has so far been gathered on the factors that contribute to the success of innovation in information technology. A number of studies focused on the innovations in either hardware (e.g. Iansiti & Clark, 1994) or software development (e.g. Cusumano & Selby, 1995). No study has taken an integrated perspective of the ICT sector, taking aspect into account both services and products. Furthermore the focus has been on identifying factors.

Eisenhardt & Tabrizi (1995) examined the computer industry in Japan, Europe and the United States by looking at 72 products in 36 computer firms. The focus was on PC, workstation, mainframe, and peripheral segments in the hardware sector. Performance was measured in terms of speed of development. They contrasted a 'compression model' of innovation (compressing sequential development stages) with an 'experiential model' (characterised by flexibility of design and improvisation). They concluded that product teams developed ICT products more quickly when acting in an experiential and improvisational design stage. Multiple design iterations, forceful project leadership, and multifunctional teams accelerate the innovation process. In contrast with expectations about tools to improve rapid application development, the use of computer-aided design (CAD) systems, schedule attainment planning and rewarding, supplier involvement and parallel development stages and pre-development planning failed to accelerate the development. Even more remarkably, this mechanism could even slow down the innovation process. In summary, the key outcomes of Eisenhardt & Tabrizi are the following:

- cross-functional teams;
- iterative prototype and test process;
- limited planning and limited use of CAD;
- no rewards for employees when reaching schedule deadlines.

Iansiti & Clark (1994) performed a variance study in the mainframe sector investigating the capability building process and the effect on dynamic performance. They found evidence that the capacity to integrate the knowledge (bases) in an organisation through effective problem solving is conducive to performance. In addition, Iansiti & Khanna (1995) presented a longitudinal study on the technological evolution in the mainframe computer industry. They presented a critical path model to investigate the influence on product development of complexity, rapid technological change and substantial discontinuities in IT components. The study showed that the ICT sector is characterised by obsolescence of technological capabilities. Iansiti *et al.* (1994, 1995) argue that firms need to develop dynamic capabilities to constantly renew the knowledge in the organisation.

Voss (1988) investigated the key success factors in new product development at the company level of 16 application software packages. The study resulted in seven key determinants of success and failure in software innovation processes. Success was determined at three levels: installation success, commercial success and a composite measure of overall success. Table 2.3 reports the set of critical success factors. External communication is important to obtain a clear view of the industry context relevant to the software-operating environment and better identification of potential markets. Efficient development work is reflected in lower costs and less lead-time. This is associated with proper planning and pre-development work. Project championing as well as sufficient resource commitment to innovation represented the top management priority. Priority was measured in terms of resource availability. Product advantage in terms of software functionality revealed to be important in the diffusion of a software innovation. Reference sites, demonstrations, in-house or with independent users, provided the credibility for the product and the confidence for potential customers. A culture stimulating risk taking and accepting uncertainty were beneficial to the success of the software development.

Table 2.3: Key success factors in software development (Source: Voss, 1988)

- active communication with industry of application
- efficient development work
- short development lead time
- project champion during innovation
- project champion during commercialisation
- resources committed (%)
- quality of demonstrable capability
- quality of reference sites
- risk-taking climate

An additional study on the success of collaborative product development (Bruce *et al.*, 1995) investigated the collaborative ventures of 100 ICT companies in the United Kingdom. The driving forces behind collaboration are the reduction of time-to-market and the sharing of costs and risks in product development. Remarkably, the study revealed that the expected benefits may not be accrued and collaboration even has a negative impact, e.g. longer lead times, higher costs and an inefficient control. The study reported that difficulties in managing external collaboration relate to the paradoxical or 'extraordinary management' requirements (Stacey, 1996; Den Hertog & Huizenga, 2000). Such paradoxes appeared between:

- starting co-operation with potential competitors and continuing to be competitive;
- retaining flexibility while establishing goals and project plans;
- balancing the management focus on the internal and external factors that influence the commercial outcome.

We now turn to a discussion on the use of the KSF concept from a more theoretical standpoint.

2.4   INNOVATION RESEARCH STREAMS: A REVIEW

The multidisciplinary nature of innovation research has enriched the understanding of what distinguishes the better performing innovators. Let us briefly address the KSF approaches and methodology. The range of key success factors point to a small set of strategy, process, organisational and environmental factors. However, as Brown & Eisenhardt (1995) argue, there is a lack of theoretical understanding on success factors. Despite its multidisciplinary character, this research field has not provided a theoretical base and deeper understanding on each of the key success factors. Furthermore, little is known about the explanation behind and the relations between the factors. Brown & Eisenhardt (1995) structured the research field and identified the work by Cooper and similar scholars as a research stream, which they refer to as the 'rational plan'. This stream characterises a large part of the research on innovation. Their classification (table 2.4), which complements our earlier distinction between process and variance research, includes the following:

- product development as a *rational plan*: containing research on factors determining financial performance of product success. Project planning, cross-functional teams and management commitment are often regarded as key success factors;

- innovation as a *communication network*: research looking at the internal and external communication by project teams in product development process;
- innovation as *disciplined problem solving*: research concentrating on the problem-solving capacity of the project team and the heavyweight project leader and the importance of top management commitment.

Table 2.4: Innovation research streams (adapted from Brown & Eisenhardt, 1995, p. 347)

Concepts	Rational plan	Communication web	Disciplined problem solving
Key idea	Product advantages market attractiveness and rational organisation are conducive to success	Success via internal and external communication	Success via problem solving
Theory	Non theory	Information processing	Information and resource dependence
Methods	Surveys Bivariate analysis One information source dependent and independent variables	Deductive and inductive reasoning multiple sources multivariate analysis	Multiple sources case-based research from inductive to deductive
Performance	Financial success in terms of profits, sales and market share	Perceptual success by team and management ratings	Operational success in terms of productivity, increase and time to market
Example study	Cooper (1996)	Ancona & Caldwell (1992)	Takeuchi & Nonaka (1986)

Studies belonging to the rational plan (Cooper, 1996; Maidique & Zirger, 1985; Rothwell, 1972) analyse the characteristics of successfully innovating firms and compare new product success and failure. The communication web research stream is in contrast with success factor studies for its linkages to information processing-based theories. The research stream of disciplined problem solving is more concerned with the organisation of teams and the role of team leadership.

Major contributions and critiques
Rational plan research is a domain characterised by its empirical nature, variety of methodologies, focus on financial success measures and lack of integration of earlier findings. There exist studies that rely on qualitative data gathering techniques (Takeuchi & Nonaka, 1986), comparative case studies analysis to develop theory (Bourgeois & Eisenhardt, 1988), or large scale surveys (Cooper & Kleinschmidt, 1995; Calantone & Di Benedetto, 1993). The field is dominated by empirical studies and is primarily based on assessing the correlation between the dependent variable success and the independent variables. What is typical of KSF studies? The empirical research is highly exploratory in nature and lacks synthesis (Montoya-Weiss & Calantone, 1994). Due to variation in concepts, variable definitions, research contexts and operationalisations, there is a pluriform collection of studies. Despite these critical notes, these studies have made some clear contributions (Montoya-Weiss & Calantone, 1995).

- The identification of a core set of key factors that contributes to innovation success. There is remarkable consistency of some of the results, despite the high variance in data sets, methodologies and operationalisations. This is due to numerous differences in sample size, the scope of the study (product, project, programme or firm level), the level of data collection, the performance definition, and the sector focus of the study. This makes generalisation and integration of conclusions difficult.
- The studies concentrate on identifying rather than explaining success factors and are primarily concerned with applied research. But this has helped managers to better understand the management of innovation.
- The retrospective and replication character of studies has strengthened the external validity. But many studies incorporate methodological threats on the internal validity due to non-randomisation of project cases and a lack of explicit selection criteria.

Valuable research directions
There is a blank spot on the empirical research map of studies incorporating both the strategy and organisation factors. A great deal of success factors has been studied at the project or R&D programme level or analysed from one functional angle. However, success at the company level is different. Success/failure studies have used numerous metrics to assess the product success or project success. But innovation performance at the firm level

involves more and complex factors. At this point the research domain faces intriguing challenges. Cooper & Kleinschmidt (1995, 1996) argue for a more macro view of innovation by conducting firm level innovation research, because of:

- the unit of analysis: success at the firm level is different from project level success;
- the mis-specification of factors: company practices can prevail that are not observed nor measured at the project level but are important to success;
- the limits inherent to the research design: when analysing project success and failure, firm characteristics do not differentiate when both project success and failure are subject of analysis. Firm characteristics will be common to both projects.

An innovation strategy that is not explicitly stated might, for example, play a role as an underlying factor explaining success. This notion of underlying factors may be of little relevance to comparing one new product success and failure, but can be essential when regarding a whole range of innovation projects, e.g. in successive time periods. More firm level features can be of vital importance to innovation performance. In particular strategy, organisation and human resource management are structural characteristics that can influence the performance in the longer run. Similarly, Brown & Eisenhardt (1995) encourage research into 'the primary relations of innovation'. In other words, analyse the causal relations between performance, organisation and strategy. They support more firm level research to account for the role of company-wide competencies that operate at the project level of innovation. These are not identified nor measured when the project is the unit of analysis. Furthermore, they encourage research that develops insight into the managerial role in development processes. Management concepts such as strategic vision and management control are still vague in this research field. According to Rothwell (1994) more insight is needed into the qualities and competencies that control and steer innovation processes.

At the same time, numerous studies have reported the effects of strategic factors upon innovation performance. Especially product advantage, technological synergy and marketing synergy are among the frequently mentioned factors that are tested and found to be critical (Montoya-Weiss & Calantone, 1994). Yet when focusing on the question of 'why strategy is relevant' the innovation research field misses solid findings. Therefore more research can bridge the gap between the innovation literature and the competitive strategy literature. This could result in a more profound role for strategy in new product development. This is in line with Mintzberg (1994), who remarked the recent theoretical debates in the field of strategy and the remarkable rise of 'strategic thinking in competencies and resources' and fall of 'strategy as planning and positioning'. Conducting more strategy-oriented research implies something different than merely testing another list of strategic elements. It involves both the incorporation of the theoretical and empirical findings in the strategy field and the understanding of new developments regarding resource-based and competence-based approaches to strategy (Teece *et al.*, 1997; Barney, 1991). Or as Montoya-Weiss & Calantone argue (1994, p. 413):

> [C]ontinued replication of the dominant strategy factors will not answer any new substantial questions, nor will it contribute to progress in this field of research unless important moderator variables, study characteristics or methodological issues are being addressed.

Innovation strategy

According to some studies (Dwyer, 1991) strategy is the essential cornerstone of innovation success. A business strategy is actually the outlet of a strategic vision about future markets and technologies. An explicit resource commitment to innovation and the presence of a shared vision about new productmarket combinations are critical components in a business strategy (Dwyer, 1991). This implies that the innovation strategy seems like a formal arrangement for the resources, which might raise several questions. But what about the strategic priority setting? Do company objectives on sales, profits and market shares conflict with innovation? And what about timing as a strategy versus product differentiation? Should companies develop areas of product advantage that support a differentiation strategy?

Shrivastava & Souder's key success factor study (1987) argues that that the new product strategy should be explicitly part of the business strategy. Cooper & Kleinschmidt (1995) strongly argue for the strategic need for product advantage and a separate product strategy. Other studies (Roussel *et al.*, 1991; Rothwell, 1992) provide arguments for the selection and composition of project portfolios, and the screening of new product ideas and technologies to assess the market attractiveness. There clearly seems to be a dispute in this part of the empirical literature.

On this discussion issue, Barczak (1995) conducted a study with 128 firms in the telecommunications industry. Three new product development strategies (NPD) were compared: a first-to-market, a fast follower, and a delayed entrant strategy. Remarkably, this study found no support that a single NPD strategy stands out as being better that any other. The results indicated that a company's focus should be on ensuring the best possible fit between the chosen NPD strategy, the corporate goals and the capabilities. Besides this specific outcome, the conclusions include the following:

- Performance is related to the use of cross-functional teams (e.g. project and R&D teams), a finding consistent with the Cooper & Kleinschmidt and Clark & Wheelwright studies. Barczak's study suggests that first-to-market firms rely more on R&D teams than fast followers or delayed entrants.
- Product champions, managers that personally commit to, guard and invest in an innovation project, are valuable contributors and influence the success of new product programs. This finding occurred in other studies (e.g. Allen (1978) on R&D professionals as technology gatekeepers) and confirms prior research on champions and new product success factors (Stacey, 1996; Gobeli & Brown, 1988).
- Remarkable is the emphasis on the importance of idea generation and idea screening for success. Prior research already identified the importance of pre-development activities. In this study both tasks have been identified as critical activities, for all three NPD strategies. In addition, the results indicated that a fast follower should offer distinctive products. This implies that fast followers should emphasise on such activities as concept definition and (consumer prototype) testing.

Let's take a closer look at the innovation perspective in the recent strategy literature.

Table 2.5: Overview of a selection of studies of innovation key success factors

Sources	Listed key success factors	Research context
Maidique & Zirger (1985)	• Product advantage, market attractiveness and organisation	70 success-failure product pairs in 21 case studies at the company level; a longitudinal study in the US electronics industry

Cooper & Kleinschmidt (1987)	• Product advantage
	• Organisation of the internal development process
	• The pre-development planning
	• Top management commitment to the project
	• Efficiency and planning execution R&D process
	• Strong internal communication between groups
	• Fit between production, research, sales and marketing requirements of the new product and the available resources and skills (synergy)
	• Market conditions:
	  - markets characterised by growth and relative weak competition
	  - markets characterised by a high degree of change in customer needs for a product category
		203 products, of which 123 were a success and 80 were failures in 125 manufacturing companies

Shrivastava & Souder (1987); Griffin & Hauser (1996); Johne & Snelson (1988); Song & Parry (1996, 1999)	• The firm's new product development process and the specific activities within this process • The new product strategy as part of the business strategy • The way projects are organised • The firm's internal culture and climate for innovation • Management commitment to new product development	Various performance measures, various industries

Cooper & Kleinschmidt (NewProd studies 1979, 1993, 1995)	• *Strategic factors* Product advantage, technological synergy, marketing synergy, resource capacity, product strategy • *Development process factors* Protocol, quality of technical and marketing competencies, quality of pre-development activities, top management support & skill, financial and business analysis, time-to-market, costs • *Organisational factors* Internal and external relations, teamwork, communication • *Market environment factors* Market potential, environment, market competitiveness	Financial performance and management ratings
Stacey (1996); Clark & Wheelwright (1993); Gobeli & Brown (1988)	• Multidisciplinary project teams and heavyweight project leaders	Multiple sectors
Cooper & Kleinschmidt (1993)	• Product advantage and pre-development work, • Market attractiveness not found to be a key success factor	Chemical industry
Barczak (1995)	• Fit between strategy, corporate goals and capabilities, instead of new product development timing strategy • Cross-functional teams • Product champions • Idea generation and idea screening	High or low performance of 128 companies in telecommunications industry classified as: first to market, fast follower, delayed entrant
Gersick (1994)	• Milestones are an effective tool for advancement in conditions of rapid technological change	

Sources	Listed key success factors	Research context
Cooper & Kleinschmidt (1995, 1996)	• Strong project management • Quality of project management included an emphasis on up-front homework, a sharply defined early product definition before development, tough milestones and flexibility to combine stages and decision points when needed • The role of pre-development work includes a detailed market and technical assessment	Success at company level
Ancona & Caldwell (1992); Gersick, (1994)	• External communication: ambassadorial and task co-ordination • Internal communication: concrete project plans with deadlines and milestones are associated with motivation of team members	45 product development teams, 5 high tech company case studies
Damanpour (1991)	• List of organisational dimensions that moderate innovation performance	Meta-analysis of innovation studies Orientated at effective use of information and acceleration of process steps
Montoya-Weiss & Calantone (1994)	• The nature of problem solving and the interfaces between organisational members. • Use of cross-functional teams, • Cross-functional interfaces between departments, • Parallel rather than sequential development • Shared responsibilities	Innovation performance and shorten the time-to-market
Brown & Eisenhardt (1995)	• *Research stream: rational plan* Project planning, cross functional teams and management commitment	Meta-analysis. Financial success in terms of profit, sales and market share

Sources	Listed key success factors	Research context
Brown & Eisenhardt (1995)	• *research stream: disciplined problem solving* Problem-solving capacity of the project team, the heavyweight project leader and the importance of top management commitment	Meta-analysis. Operational success in terms of productivity increase and time-to-market
Voss (1988)	• Active communication with industry of application • Efficient development work • Short development lead time • Project champion during innovation • Project champion during commercialisation • Resources committed (%) • Quality of demonstrable capability • Quality of reference sites • Risk-taking climate	16 product development projects in software sector
Eisenhardt & Tabrizi (1995)	• Product teams acting in an experiential and improvisational design stage • Multiple design iterations, • Forceful project leadership, and • Multifunctional teams accelerate the innovation process. In contrast with expectations, the following factors showed to be ineffective in accelerating the development: • The use of computer-aided design (CAD) systems, • Schedule attainment planning and rewarding, • Supplier involvement and • Parallel development stages • Pre-development planning	72 development projects in 36 computer firms, speed of innovation process Used as success variable

3. The Nature of Strategy

3.1 INTRODUCTION

Innovation research used to be predominantly concerned with the key success factors in the design and development of novel products (e.g. Dwyer, 1991). An important key success factor is strategy, in particular new product strategy. However, the KSF literature does not go into detail about the content of strategy. By focusing on the question 'why strategy is relevant' the innovation research field lacks solid findings. Empirical research (e.g. Barczak, 1995; Cooper & Kleinschmidt, 1995; Roussel *et al.*, 1991; Rothwell, 1992) on new product development strategies (NPD) has shown that a company should focus on the best possible fit between the chosen NPD strategy, the corporate goals and the capabilities. Others argue that each company's situation is determined by its history (e.g. Teece's *et al.* (1994) 'path dependency') and the current competitive strategy. This part of the empirical literature seems to be subject to debate.

This might raise numerous questions. What is a successful new business strategy? What about the strategic priority setting? Do company objectives regarding sales, profits and market shares conflict with innovation? What about timing as a strategy versus product differentiation? Should companies develop competencies? In this chapter, we will try to address these issues on the basis of the strategic management literature, in order to contribute to the knowledge development of innovation.

But why do we have to look at the role of strategy for innovation management in the ICT sector? We assume that innovation is the source of business value, growth and competitive edge for ICT companies. The ICT sector is a young, dynamic high-tech sector, which drives on developing new business. We are interested in observing how firms differ in their strategy. The contrast between empirical insights into strategy of ICT firms and recent descriptive theory will contribute to our understanding of the future directions of the ICT sector. According to Hamel & Prahalad (1994) and other scholars (e.g. Den Hertog & Huizenga, 2000; Teece *et al.*, 1997;

Quinn, 1993; Sanchez & Heene, 1997) strategy deals with answering both the two following questions:

- 'Are we doing the right things', resulting in an answer that must be provided for the firm to continue to exist
- 'Are we doing the *right* things right?', resulting in an answer, if positive, that adds to the firm's differentiation and competitive edge.

There is a strong element of corporate strategy in this, with different firms taking different decisions on potential clients and new services and products. A large number of the firms in the ICT sector might have problems in developing such an innovation strategy.

Furthermore, in many sectors, and especially the ICT sector, competition is shifting away from producing towards serving customers. Businesses increasingly recognise that the nature of their offering involves a mixture of tangible and intangible properties. These service dimensions enter into all products and businesses, which raises new challenges for strategy. The competitive environment becomes more complex due to the rise of the 'service intensity' of all markets. Consequently, ICT firms need to consider this development in their strategy. In the contemporary strategy research literature, new perspectives appear that describe the role of such 'complementary assets', e.g. the dynamic capabilities of Teece *et al.* (1994).

3.2 STRATEGY SAFARI: AN OVERVIEW OF SCHOOLS OF THOUGHT

In this section we briefly discuss the main characteristics, the contributions and the critiques on the strategy schools and the innovation focus. Mintzberg (1994) argued that a first glance of the literature shows that the strategy field is characterised by fragmented unconnected theoretical perspectives. Basically, two opposite perspectives have emerged: the behavioural and organisational perspective on the one hand, and the economic perspective on the other. The former research stream concentrates on the internal components of the firm while the latter analyses the firms with an external focus on competition. This has led to a theoretical polarisation and fragmentation of strategic concepts. The research field has been dominated by a view of strategy formation that is referred to by Whittington (1993) as the 'classical approach', by Nonaka & Takeuchi

(1996) as the 'structural approach', and by Chaffee (1985) as the 'linear model of strategic management'. These early approaches are characterised by a formulation of strategy as a rational process, prescribed by techniques to identify current strategy, analyse the environments, the resources and revealing strategic alternatives (Pettigrew, 1985, p. 276). Sanchez and Heene (1997) argue that the traditional strategy lost much of its power to guide the management of organisations in today's business, due to the dynamic competition in many industries.

The design, planning and positioning school
Mintzberg (1994) identifies nine different perspectives of strategic management (for a detailed description of all the characteristics of the schools we refer to the article). The most interesting perspectives include the design school, planning school, the positioning school and configuration school. The configuration school will be discussed in chapter 4. Chandler (1962) has initiated the strategy structure debate, suggesting selective strategic configurations.

The design and planning perspectives provide an overview of the state of the field from the 1970s until the late 1980s. Both the design school and planning school regarded strategy formation as a controlled, conscious process of thought. Characteristic of this strategy perspective is that responsibility rests with the chief executive officer, which was regarded '*the strategist*'. Strategies should be explicit, thus requiring a simple model of strategy formulation. Implementation was seen as a sequential, separate task, executed after the formulation stage. Both perspectives built upon the work by Andrews (1971) and Ansoff (1965). This strategy thinking was popularised by SWOT analyses examining the strengths and weaknesses of the firm and the opportunities and threats from the environment.

The ideas and concepts of strategy formation as a conceptual and formal process were challenged and questioned. The planning tools were intuitively sound, but no further insights existed into how to assess strengths and opportunities. The research field was dissatisfied with the prescriptive attitude of the strategy schools (Collis & Montgomery, 1995; Mintzberg, 1994). The main criticism was related to the relevance of the strategy design & planning model, which was inappropriate or even counterproductive for organisations. In addition, the disconnection between formulation and execution implied that strategic thinking was performed in the first part only. Adaptation of a strategy and learning from previous strategic formation activities was neglected in both perspectives. Also, the idea of one single individual as a strategist was far from business reality. Overall, the assumptions from the design and planning school were likely to lead to an oversimplification of strategy, with its inherent risk of a mismatch with the market.

The contributions of the planning school next to the design school were the development of a highly formal planning and execution process. The informal design perspective was replaced by a formal explicit sequence of steps with the CEO as *'strategic architect'* accompanied by strategic planners. Strategy implementation became a matter of developing objectives, budgets, programmes and operating plans. Although resulting in valuable contributions in terms of techniques, the perspective was criticised. Planning did not encourage change but was essentially an inflexible process favourable to a stable environment. While decomposing and formalising process steps, the strategic planning lost synthesis. Strategy became a collection of separate components without an integrative element (Donaldson, 1995).

The positioning school
With the emergence of concepts of 'competitive positioning' Porter (1980, 1985) gave new stimulus to the strategy field. Porter's work extended the 'structure – conduct – performance' framework by analysing the position of the firm with respect to the competitive environment. In his view the firm is part of an environment that is dominated by five competitive forces: the customers, competitors, potential market entrants, suppliers, and substitute products. These forces influence the strategic position of the firm, which is defined in terms of product market combinations and the fit between strengths, weakness and market opportunities. Based on this competitive forces framework a firm can pursue three different generic strategies for achieving high performance:

- Cost leadership strategy: a firm designs its internal processes more effectively than its competitors, resulting in producing at lower costs.
- Product differentiation strategy: a firm develops the ability to differentiate from its competitors by positioning its products and services differently on price, image, support, quality or design.
- Focus strategy: a firm serves a niche market segment with a focused customer base. The focus strategy can be pursued on the basis of low costs or differentiation.

Essential in Porter's analysis is the idea of the firm as a bundle of activities to deliver products and services. This idea has been developed using the concept of the value chain. The value chain describes a sequence of strategically important value activities (the 'primary process') and the secondary supportive processes. By means of defining the value chain a company can define its activities that add most value and which distinguish the firm from its rivals. Furthermore, the value chain provides understanding of the cost behaviour of activities and reveals the relation within the chain of activities, but also that between value chains of suppliers and customers.

The positioning school has also suffered from criticism. It is a strategy perspective biased by conventional well-established big firms in large industries. The role of entrepreneurship is hardly conceptualised (e.g. Stopford & Baden Fuller, 1994). Although the perspective discusses the role of market entry and exit barriers, the emphasis is positioning in current markets and less on creating future generic positions (e.g. Hamel & Prahalad, 1994).

Strategy and performance: external versus internal focus

Over the past years criticism has been concerned with the applicability, universality and validity of strategic concepts. The strategic research field was facing the challenge of understanding more deeply the complex nature of strategy, which required improvement of the practical relevance. Rumelt, Schendel and Teece (1991, 1996) argued that despite the criticism the theoretical strategy research streams have contributed to a deeper understanding of the elementary strategic questions:

- How do firms behave?
- Why are firms different?
- To what extent should structures be decentralised?
- What is the function of corporate headquarters in a multi-business firm?
- What determines the success or failure of a firm?

Especially the second question has recently led to a strong debate among strategy scholars. The industrial organisation paradigm focuses on industry effects to explain superior rents that firms can collect (Hansen & Wernerfelt, 1989; Vasconcellos & Hambrick, 1989; Rumelt, 1991; Schmalensee, 1985). A study on the variance of rates of return on assets among business units Schmalensee (1985) concluded that industry effects are an important variable explaining performance variance. This argues in favour of the focus on the industry-level analysis. However, Rumelt (1991) performed a similar study with longitudinal data, which revealed that the dispersion of returns is explained by business unit effects more than by industry effects. He concludes that industry analysis neglects the fact that industries are too heterogeneous to support one general industrial organisation theory. Furthermore, empirical studies have revealed that differences in profitability *within* an industry are more important than differences *between* industries. This suggests that innovation success is driven by strategic and resource differences between firms rather than by the industrial characteristics.

These perspectives should be complemented by an approach taking into account the internal resources and competencies that matter. Sanchez and Heene (1997, p. 303) speak of:

> A movement to make strategy theory more relevant to contemporary forms of competition by rethinking the content and process of strategic management theory and practice.

In a special issue of the *Strategic Management Journal*, Prahalad and Hamel (1994, p. 6) refer to:

> A need for a basic re-evaluation in order to pave the way for new ideas ... [and] ... many of the assumptions that were embedded in traditional strategy models may be incomplete or outdated as we approach the new competitive milieu.

What caused the theoretical debate on strategic models of competitive advantage? Much of the discussion concentrates on the rapid change of the external environment (e.g. Mahoney & Pandian, 1992; Collis, 1991; D'Aveni, 1996). A change that has evolved slowly, almost indiscernibly, but had a tremendous impact on business and competition. Prahalad & Hamel (1994, 1991) point to some major drivers for a revitalisation of approaches to strategic thinking and development, such as the need for 'competing for industry foresight' and 'strategic intent and expeditionary marketing'.

3.3 RESOURCE-BASED VIEW OF STRATEGY

Contemporary strategy research has shifted to perspectives that emphasise the resources of the firm as a productive factor and competitive weapon. The internal firm resources are regarded as a key variable for competitive advantage (Hoopes *et al.*, 2003; Teece *et al.*, 1997; Amit & Schoemaker, 1993). Basically, two opinions of resource-based perspectives can be distinguished:

- One perspectives underlines the idea that internal resources are valuable and investigates why firms need to concentrate on resource accumulation. This view is rooted in the seminal work of Penrose (1959), which stressed the importance of resources for firm growth.
- The second perspective, which is complementary to the first, underlines the idea of resource exploration and exploitation, in particular with respect to innovation.

What are the essential arguments of this particular view? The resource-based perspective regards the firm as a bundle of durable, intangible assets. Firms are assumed to be heterogeneous and differ in their collection of physical and intangible resources (Hoopes *et al.*, 2003). This resource endowment of the firm can account for a sustained competitive advantage by capturing entrepreneurial rents (Peteraf, 1993; Barney, 1991; Dierckx & Cool, 1989; Wernerfelt, 1984). Collis (1991) points out that these resources are the input for conducting functional activities and make a firm strategically different. The literature on the resource-based view (RBV) of the firm uses a variety of definitions for resources. The RBV of the firm rejects the Darwinian selection assumption of the external environment that characterises the positioning perspective. A theoretical research stream that further elaborates on this principle is the population ecology (Hannah & Freeman, 1989). In this perspective, strategy formation is a passive process. The environment is viewed as a set of strategy-dictating forces. The principle of selection and retention ensures that organisations find ecological niches and must ensure to adapt over time. If they don't, they will disappear from the market.

The value of resources
Teece *et al.* (1994) observe from a theoretical standpoint that resources derive their value from qualifying on three criteria. The first criterion refers to the 'stickiness' of resources in the *short term*. Stickiness is derived from:

- the complexity of business development;
- the difficulty in asset tradability;
- the lack of superior information on the value of a resource.

The interaction between these criteria determines the value of a resource (Collis & Montgomery, 1995). In order to be valuable in the *long term*, a resource has to pass a range of tests. Long-term value only holds for those resources that are *hard to imitate*. The difficulty of imitation may rest in geographical immobility, imperfect information or firm dependence. The accumulation over time also generates barriers to imitation. Resources that are collected over a longer period of time become *path-dependent* (Huizenga, 2002a). Competitors have to invest in resources accumulation to acquire this valuable resource. R&D investments, for example, are closely related to the idea of path dependency and inimitability (e.g. Iansiti & Clark, 1994; Pisano & Wheelwright 1995). The second criterion for the value of a resource is concerned with the *durability* of the resources. A durable resource is a resource that does not become obsolete or is not substituted in the short term.

The third market test for resource value is the *appropriability*, or rent-earning potential, of resources. It refers to the owner that actually receives the rents from a resource. For example, the control over human and technological resources depends on the relationships between the organisation and its employees, technology alliance partners and its customers. If this relationship is strongly embedded in the organisation, the management might be better able to appropriate the rent. The appropriability is strong if the resource is difficult to replicate, and if there are barriers to imitation, such as a patent system.

A number of other concepts related to the value of a resource include the stock and flow of resources. On the one hand, the *stock of resources* can account for rent earning and be conducive to competitive advantage or firm success. The efficient deployment of an existing stock of resources, e.g. a market share, a brand name, or the number of employees or the customer portfolio, can create such entrepreneurial rents. Dierckx and Cool (1989) refer in this respect to 'asset mass efficiency'. On the other hand, the *accumulative flow of resources* over time can be a determinant of superior performance. Examples are marketing investments for higher brand recognition or a training programme. Dierckx and Cool (1989) speak of 'time compression diseconomies', which are acquired by accumulating resources at an accelerated pace resulting in a higher firm performance.

Key contributions of RBV to strategy research
The main contribution from the resource-based view is the ability to explain long-lived differences in firm performance that cannot be attributed to industry differences (Peteraf, 1993). This is an interesting element for our sector-specific study because it suggests that frontrunners and pack members differ in their stock or flow of resources. This subject will be explicitly discussed in chapter 7.

The resource-based perspective has offered a competing view to the planning and positioning view of strategy by drawing attention to entrepreneurship. Preoccupation with market positioning has been supplemented by internal phenomena such as resources, endowments, investment and management systems. These factors have not attracted much attention in the positioning literature. With the arrival of the RBV there is an explicit recognition for the importance of unique managerial resources and the resource investments.

The idea of resource superiority has provoked new approaches to analysing diversification and vertical integration like mergers and acquisitions (Teece *et al.*, 1994). The decision to enter a market or to develop a new product or service rests the availability and uniqueness of the resources. The resources argument has shed new light on the make-buy decision of the firm (Stopford & Baden-Fuller, 1994). Firms invest in assets that ensure a competitive advantage. An interesting research question is whether firms should develop these assets internally or (partially) source them from the market.

Criticism on the resource-based perspective
The main criticism of the resource-based perspective is its lack of attempt to explain the nature of the isolating mechanism that provide entrepreneurial rents and competitive advantage. Teece *et al.* (1994) argue that the dynamics of competition are ignored, the resource development and exploitation is undefined, and the selection environment is not taken into account. The theoretical question is whether there is a relationship between resources and competitive advantage. This is in contrast with the idea of Wright & McMahan (1998), who argue that the stock of human resources is directly related to competitive advantage.

RBV investigates the relationship between asset specificity and firm specificity (Barney, 1986). Although the RBV addresses the essence of resource accumulation, the view has not elaborated on how the combination and co-ordination of resources occurs. The dynamic capability perspective addresses the organisational mechanisms to integrate resource flows. Wernerfelt (1984) concludes that the resource-based perspective invites the consideration of managerial strategies for developing new capabilities.

The resource-based perspective concentrates on organisational collections of resources and the relation with performance. Although the RBV implicitly acknowledges the role of exploring and exploiting resources (McGrawth, 1997) it does not formally integrate these concepts in its perspective. Resources are accumulated without expressing how this process actually takes places. Resources have to be renewed, as they do not render valuable services eternally. Resources become obsolete and have to be replaced. The resource-based view fails to account for these dynamics and does not conceptualise how resources develop. The collection of resources is a rather static concept when environmental changes are not addressed. Or as McGrawth (1997, p. 33) states:

> If exploration, search and the creation of new routines are processes left to the
> disappearance of slack and increasingly poor performance spark, the firm may be
> poorly equipped to then engage in the dissipative, discovery oriented learning
> which its new situation may require.

3.4 THE DYNAMIC CAPABILITIES PERSPECTIVE

Sanchez & Heene (1997, p. 306) point out that the resource-based view of the firm failed to build a coherent theoretical model and left the research field with conceptual gaps and fragmented theory between the behavioural and economic perspectives. A new breed of literature, inspired by management practice and strategy theory, initiated the development of an integrative strategic management approach. The dynamic capability view has the following characteristics:

- The internal perspective supplements the external analysis by emphasising current markets.
- It refers to competencies and future markets as the source of competitive advantage (Teece *et al.*, 1997; Hamel & Prahalad, 1994).
- A firm becomes successful by developing an advantage based upon the internal competencies and knowledge (Barney, 1991; Leonard-Barton, 1995; Nonaka & Takeuchi, 1996).

Sanchez & Heene (1997, p. 307) formulated this as:

The central objective in competence theory and management practice is understanding the co-evolutionary dynamics of environmental and organisational change and their roles in shaping organisational competencies.

Capability and competencies versus resource

Many discussions in strategy related journals described the essential differences between capabilities, competencies and resources (see box 3.3). Grant (1991) argued that competencies refer to the ability of a firm to attain a durable strategic advantage through sustaining the co-ordinated deployment of resources and capabilities. There is a key distinction between resources and capabilities. Resources are inputs into a transformation process. When combined, resources become productive and generate value. A capability is the capacity for a bundle of resources to conduct an activity (Grant, 1991). They are 'repeatable patterns of action in the use of assets to produce goods and services' (Sanchez, Heene & Thomas, 1996). Capabilities involve complex patterns of co-ordination and co-operation between individuals, teams and organisations. Such integration can take many forms, such as project team organisations, models of communication, intra-business relations and inter-business relations.

Key contributions

The interest for the internal strengths of the firm as a source of competitive advantage has gained a lot of attention in the literature. Whereas economic perspectives view the firm as rational agents pursuing maximising behaviour, competence theory regards the firm as an open system of knowledge asset stocks and flows (Den Hertog & Huizenga, 2000; Dierckx & Cool, 1989). The latter explicitly relates internal organisational processes with the external interfaces of the firm. This research field explicitly attempts to integrate the fragmented strategic perspectives that prevailed until the 1980s. Recent publications have proposed *the theory of competence-based competition* (Sanchez & Heene, 1997; Heene & Sanchez, 1997; Sanchez, Heene & Thomas, 1996; Hamel & Heene, 1994; Hamel & Prahalad, 1994). The propositions of this theory supplement and enrich the resource-based perspective. Lowendahl & Haanes (1997) point to the strengths of the perspective:

- its emphasis on an open holistic and systemic view of the organisation,
- its focus on the dynamics of competence building and leveraging, and

- its extension of definitions of competencies and capabilities.

Although the conceptual integration is a valuable objective for this research stream, there is the risk that the creative ideas of this paradigm are weakening. The explosion of theoretical and empirical studies has fragmented the competencies research field with respect to definitions (see box 3.1).

Box 3.1: Definitions in the dynamic capability perspective

The distinction between core competence and capabilities is not always clear. The concepts both focus on behavioural aspects of strategy and are interested in the 'why' and 'how' of performance differences. But many dimensions seem to apply to the concepts, e.g. technology integration (Hamel & Prahalad, 1994), transformation of business processes, and innovation (Leonard-Barton, 1995; Teece *et al.*, 1994; Pennings & Harianto, 1992; Dougherty, 1992). A lot of authors have analysed competencies and capabilities. This framework integrates several other definitions. Sourceable capabilities refer to Leonard-Barton's (1995) supplemental capability, Nelson & Winter's (1982) routines, and the distinctive competencies by Hitt & Ireland (1985) and Snow & Hrebiniak (1980). Core capabilities refer to Collis' (1991) organisational capabilities. Core competencies refer to Hamel & Prahalad's definition (1994) and to the firm-specific competencies by Pavitt (1991) and invisible assets (Itami, 1987).

The identification of core competencies is dependent on four key aspects (Rumelt *et al.*, 1996):
- Core competencies 'span' across business and product boundaries in the firm.
- Core competencies have a 'temporal dominance' over products in that they evolve more slowly than the products they generate.
- Core competencies arise through 'collective learning' and are enhanced through sharing and using.
- Core competencies are strategic.

According to Hamel and Prahalad (1994) competencies belong to the core of a firm:
- if they provide a unique competitive advantage for the firm,
- if they form the basis for new business development throughout the entire organisation, e.g. if they create new products and new markets, and

- if they provide added value to customers, e.g. by strengthening the market position of the customer in his business.

The concepts of paths, processes and positions

The key elements of capabilities are their embeddedness in the organisational processes and the integration of technology and knowledge flows to develop new products, services and process. Teece *et al.* (1997) identified three dimensions of dynamic capabilities: the three Ps of Processes, Positions and Paths. Teece *et al.* (1997, p. 518) state:

> The content of these processes and the opportunities that capabilities afford for developing competitive advantage at any point in time, are shaped significantly by the assets (= positions) the firm possesses (both internal and market) …[A]nd by the evolutionary path it has inherited.

The dynamic capabilities concentrate on change and the creation of new products, markets and technological opportunities. Dynamic capabilities focus on (Teece, Pisano & Shuen, 1997, p. 515):

> [T]he innovative successful companies [that] distinguish by timely responsiveness, rapid and flexible product innovation integrated with the management capability to co-ordinate and redeploy competencies effectively.

If a firm wants to develop and exploit its capabilities it must have a system to manage the business activities and knowledge flows (Leonard-Barton, 1995).

The *path dependency* of the firm is an important element as it restricts the resources available for new opportunities. Past strategic and organisational choices have determined the building and leveraging of the firm's durable competencies. This implies that the firm develops an organisational inertia, which limits the firm in picking up technological or market opportunities. Paths refer to the strategic alternatives available to the firm. In the learning organisation routines for innovation, knowledge development and customer services render the firm path-dependent. The resulting danger is that the firm ends up in a competition trap and is unable to catch up with new emerging technologies and dominant designs.

Table 3.1 provides an overview of the major distinction between the preceding strategy frameworks and the emerging *resource-based views*. The major contribution of the RBV perspective is its emphasis on the entrepreneurial role of the firm in the competitive environment and the focus on the internal strengths as a source of competitive advantage. The main contribution of the RBV is its ability to explain long-lived differences in firm performance that cannot be attributed to differences in industry conditions (Peteraf, 1993). The dynamic capability framework concentrates on change and the creation of new market and technological opportunities. The perspective focuses on '... the how and why question about firms building competitive advantage in regimes of rapid change ...' (Peteraf, 1993). The strategic planning and positioning schools have not addressed these questions.

Table 3.1: A comparison of the strategy schools

Strategy dimensions	Competitive positioning	Dynamic capabilities and resource-based strategy
Message	• Strategy as fit	• Strategy as stretch
Competitive environment	• Match with external requirements	• Ambitious strategic vision and leverage of scarce resources
View of the firm	• Portfolio of product market combination. Firm is part of complex competitive system	• Firm as a portfolio of resources and dynamic capabilities. Firm acts as entrepreneurial unit
Major business strategy	• Generic strategy: cost leadership, differentiation, focus	• Create new markets and build new competencies
View of innovation	• Technology development as support activity	• Primary process and source of sustainable competitive advantage
Focus of strategy formulation	• Focus upon core business activities in the value chain	• Focus on three Ps: processes, positions and paths

4. The Organisation of Innovation

4.1 INTRODUCTION

In the literature on key success factors there is a prominent place for organisational factors. The importance of the organisation is a key attribute of successful innovating firms. In fact, in the theoretical discussion on the organisation process behind innovation, scholars agree that the type of organisation is an indispensable building block. Also, most empirical studies on new product and process development draw attention to the organisation's impact on innovation success. There seems to be little controversy in the literature as to these viewpoints. On the other hand, the organisation literature is known for its rich variety in approaches, perspectives, and theories. The literature journey of this chapter is not intended to address all similarities and differences between different organisational theories. This chapter shows the variety of views of organisation on innovation. Additionally, it provides expectations of topics that are supposed to be of great importance to a knowledge-intensive business as the ICT sector.

4.2 ORGANISATIONAL PROCESS MODELS OF INNOVATION

The importance of the organisation of innovation process has been given considerable attention in the literature (e.g. Carter & Williams, 1957; Allen & Cohen, 1969; Rothwell *et al.*, 1974; Cooper, 1979; Henderson & Clark, 1990; Takeuchi & Nonaka, 1986). A great number of models of new product development processes have been proposed to address the 'how' and 'what' questions of innovation. Also, these models have been adopted by companies, which generally had a strong positive impact on a firm's innovation efforts (Cooper, 1994). Typically, each model in the literature represented an effort to describe the 'one best process model' for innovation. This attempt to describe innovation process models can be characterised as a search for the 'one size fits all' model, irrespective of the type of product or organisation. From an organisation design perspective, three generations of process models are identified (Cooper, 1994; Rothwell, 1994; Sarren, 1984). Although the ICT sector is still a relatively young industry, innovation is a key to competition and survival. We expect the later generations of process models to be relevant to the complex innovation processes taking place in ICT companies.

First-generation models: linear and sequential process 1950s-1970s
The early insights into innovation models were rooted in ideas of systems thinking in the 1950s. Systems analysis is a methodology for systematically organising large, complex processes and its constituent successive stages of activities. In the systems perspective, innovation was regarded as a co-ordinated set of procedures for making rational choices in a complex situation. The guiding principles underlying systems thinking are:

- rationality;
- linearity and predictability of activities;
- structuring of alternatives;
- a scientific-based selection approach.

Systems thinking and early work by Carter & Williams (1957) and Myers & Marquis (1969) contributed to the first generation of linear models of innovation (Cooper, 1996). These models suggest that innovation proceeds along an orderly, predictable sequential way in an attempt to minimise the risks. The linear model assumes a step-by-step sequence of independent phases. An activity was completed and the outcome was 'handed over' to the next department or activity similar to a relay race. To give an idea of the broad spectrum of linear innovation process models, Sarren (1984) identified several distinct classes of innovation models (figure: 4.1 to 4.4).

- the departmental stage models, describing the sequence of functional departments that are active in an innovation process;
- the activity stage models, showing the sequence of activities conducted in an innovation process;
- the linked department/activity models, illustrating the sequence of functional departments and their activities involved in an innovation process;
- the decision stage models, addressing the key decision moments between successive stages in an innovation process;
- the conversion process models, addressing the input and output generated in an innovation process which is regarded as a conversion process;

- the response models, describing how the organisation reacts to stimuli and the firm's response to change (e.g. a series of four stages: perception – search – evaluation – response).

Booz, Allen & Hamilton (1968) popularised the relay idea with an empirically based model (figure 4.2) that begins with an exploration of the marketplace for customer needs. Later models introduced a 'review phase' upon completion of an activity. In the 'phased review process models' each activity was followed by a formal go/no go decision.

Figure 4.1: Example of a departmental stage model (Sarren, 1984)

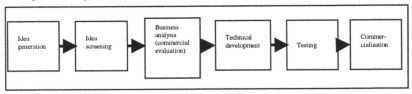

Figure 4.2: Example of an activity stage model (Based on original classification by Booz-Allen & Hamilton)

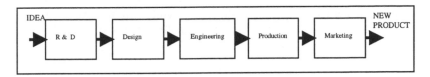

Figure 4.3: Example of a decision stage model (adapted from: Cooper & Moore, 1979)

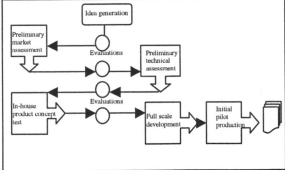

Figure 4.4: Example of a conversion model (adapted from: Twiss, 1980)

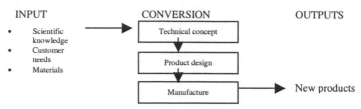

Hedging innovation

Each of the simplistic models (Rothwell, 1992) emphasises a distinct feature of the innovation process. The first three models concentrate primarily on hedging the functional departments or functional activities involved in innovation. The latter three models focus on hedging the sequence of decisions and transformations.

An advantage of representing a linear process is the potentially better control by hedging activities in the complex process of changes. Among the disadvantages are the implied rigidity of steps and the unclearness about responsibilities between successive steps. Also characteristic of these models is the absence of feedback or feed-forward linkages, e.g. when corrective actions need to be taken upstream or downstream the process.

New practical evidence (e.g. Imai, Nonaka & Takeuchi, 1985; Womack, Jones & Roos, 1990; Clark & Fujimoto, 1991) has emerged showing other shortcomings and problems of managing innovation as a linear process. The main difficulty arose in managing the interface moments, especially the linkage between R&D and marketing. At each interface, and at each decision point every idea, concept or product feature was specified in detail before being handed over. But this assumption did not seem to be valid in real life. Such a transition from one stage to another can be compared to a relay race. At critical moments, handing over the stick ('throughput') to the next stage failed. In the relay race a lot of time was wasted on such hand-over moments. The interfaces could thus become the bottleneck in each innovation stage with the ultimate consequence of delay. There appeared to be a need for streamlining activities to better organise the part-whole relation of activities.

Second-generation models: towards parallel, overlapping phases 1970s-1990s

The chain-linked integration model (Kline, 1985) is one of the popular models that emerged. This model distinguishes from first-generation models as to:

- the initiator for innovation;
- the role and presence of R&D and knowledge;
- the definition of several feedback loops and information links.

The chain-linked model consists of five phases (see figure 4.5). In contrast to linear models, this process is not initiated by research (technology push), but by market findings and invention. Research is a separate activity that is not directly coupled to the invention and innovation phases. Between these phases and the research layer exists a 'body of knowledge' already present within the company. This body of knowledge accumulates over time through new research and experience. During the different phases in the process, the knowledge layer is used to support the chain of invention, design, development and production. If this layer of knowledge falls short in providing solutions.

Figure 4.5: The chain-linked model (arrows refer to feedback loops) (Source: Kline, 1985)

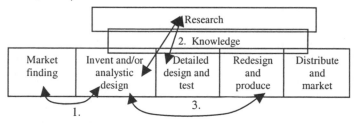

The chain-linked model represents a one-way flow from new market ideas to the distribution of a product. Essential for the working of a research and knowledge layer are the feedback loops at three levels:

- the iterative feedback between each successive phase;
- the feedback links between research, knowledge and design;
- the feedback links between upstream and downstream activities (e.g. design and produce).

The chain-linked model still bears elements of a linear sequence of process phases. The next innovation models appeared in the early 1980s. These models perceived innovation as a 'stage gate process' conducted by a team instead of functional departments (Cooper, 1996). The team jointly completes each predetermined innovation stage and proceeds to the next gate. Particularly important is the presence of overlapping activities. Working in teams proved to be a remedy for the drawbacks of sequential development. Innovation was no longer the sole task of one functional department, followed by the next department. Innovation came to involve more functional disciplines at the same time. In this perspective, innovation much resembles a rugby approach, where team players jointly proceed forward to the end. The main differences between a relay race and rugby approach in terms of innovation is the presence of overlapping development stages, the early co-operation of various functional disciplines even at the pre-development stage. Table 4.1 summarises and contrasts the generation models.

Table 4.1: Differences between innovation process models

First-generation models	Second-generation models	Third-generation models
Linear sequence of stages: activities, departments, decisions ('phased review model')	Process includes feedback loops ('chain-linked model') or overlapping phases	Process includes overlapping phases Directed at shortening time-to-market sharper product definition
Innovation initiated by R&D	Innovation initiated by market findings	Emphasis on external input from lead users, suppliers at idea phase
Technology push	Market pull and technology push	Market pull and technology push and early involvement of lead users & suppliers
No feedback/ feed-forward	Feedback and feed-forward	Especially feedback and feed-forward in pre-development stage Attention for R&D-marketing link
Mono-discipline	Cross-disciplinary between departments at successive stages (marketing and manufacturing integral part of process)	Multidisciplinary teams in pre-development stage, followed by functional groups that operate in parallel. Focus on teamlearning and subtle team control
Relay race	Rugby approach	American football
Carter & Williams (1957)	Cooper (1996), Kline (1985)	Takeuchi & Nonaka (1986)

Third-generation models: shorter time-to-market 1990s

Authors like D'Aveni (1996), Stalk (1988), Hamel & Prahalad (1994) and Teece *et al.* (1997) have emphasised a changing competitive environment. To compete, firms must continuously improve their speed to market. Shorter time-to-market within the innovation process becomes a way for firms to achieve a competitive advantage (Karlsson & Ahlstrom, 1996). Cooper (1994) refers to this movement as follows:

> Its particular emphasis is on efficiency: on speeding up an already effective second generation stage gate process and a more efficient allocation of development resources.

What are the implications for the organisation of the innovation process? Some of the problems in second-generation process models could weaken the rapid completion of an innovation process, such as:

- the delay for innovation projects that must wait to pass a decision gate;
- the lack of prioritisation in activities in the whole process;
- the delay resulting from having to pass many gates and stages;
- a tendency of processes becoming too bureaucratic.

A third generation of models appeared, driven by the quest for speed and overlapping stages. Central elements in the models are the strong upstream and downstream relations in the innovation process, cross-functional teams, and the integration of functional tasks. Case studies of NEC, Fuji-Xerox, Honda and studies in the automobile sector, which typified mass-market industry, contributed to this development of process models (Imai, Nonaka & Takeuchi, 1985; Womack, Jones & Roos, 1990; Clark & Fujimoto, 1991). These studies revealed new insights including:

- the relevance of external sources for new ideas and the early involvement of 'lead users' (Von Hippel, 1986; Thomke & Von Hippel, 2002) and specialised suppliers in the innovation process;
- the importance of internal information flows, especially the functional interface between R&D and marketing (Griffin & Hauser, 1996; Song & Parry, 1996, 1999; Souder, 1987);
- the role of tacit knowledge, team learning and subtle team control in product development (Nonaka & Takeuchi, 1996).

More recently (Nonaka & Takeuchi, 1996), the use of multidisciplinary teams was refined on the basis of case studies at Nissan in Japan. There, multidisciplinary teams were predominantly used in the pre-development stage. Once the design of the product was clear the team split up and the activities continued to be performed by functional groups working in parallel on parts of the work. The groups ensured sufficient functional knowledge input from the various functional disciplines in the pre-development stage. This method resembles the American football game. In American football, teamwork and individual actions go hand in hand within a pre-defined set of rules.

4.3   ORGANISATION THEORY: CONTINGENCY PERSPECTIVE

Daft & Lewin (Special Issue *Organisation Science*, 1993) expressed their concern about new contributions when asking:

> [W]here are the new theories of organisation …???

In this chapter we will pick up two interesting schools of thought in organisation theory that address innovation. Both schools can help us at a later stage to explain some of the typical innovation issues in the ICT sector.

Contingency approach
The dominant and leading thoughts on organisational structure were dominated by classical management theory (e.g. Taylor, 1911; Fayol, 1949 in: Shadritz & Ott, 1987). This theory was driven by principles arguing for one universal best way to structure an organisation. The development of contingency theory is a reaction to the ideas of 'one best way' to organise and manage a firm (Stacey, 1996; Lawrence & Lorsch, 1967; Child, 1984). Chandler (1962) opposed to this universality of 'one best structure', by stating that organisations adapt their structure to reach a fit between changing factors, as the world seems more complex due to different tasks and technologies in an organisation. The theory further proposes that organisations actively change to correct a structure misfit to improve performance. Empirical research (Burns & Stalker, 1961; Woodward, 1965; Lawrence & Lorsch, 1967) tried to show that success was not correlated with a simple single set of factors. Instead, it was argued that the effectiveness of a particular organisation is dependent upon a number of

factors. Donaldson (1995, p. 12) summarises this approach to organising as follows:

> Contingency theory regards the design of an effective organisation as necessarily having to be adapted to cope with the 'contingencies' which derive from the circumstances of the environment, in which the firm is operating.

Contingency factors
Research in the contingency school concentrated on explaining the degree of centralisation, which depends on contingent factors such as:

- the strategy (Chandler, 1962);
- the operational technology (Woodward, 1965);
- the rate of environmental change (Burns & Stalker, 1961; Lawrence & Lorsch, 1967);
- the firm's size (Pugh *et al.*, 1969);
- human resource (e.g. Gomez-Mejia & Balkin, 1989).

Size as a contingency factor
It was Weber's (1949) type of 'bureaucracy' that exemplified the contingency element of size to the organisational structure. In this perception an organisation consists of functional departments (e.g. marketing, production, R&D) which are supervised by higher levels. Essentially, bureaucracy is an organisational structure characterised by three elements (Donaldson, 1995, p. 39):

- functional specialisation;
- standardisation;
- formalisation.

Empirical studies (e.g. Child, 1984; Khandwalla, 1973) have shown that a larger organisation is often accompanied by higher degrees of bureaucracy. These studies support the contingency view that size leads to both functional specialisation, standardisation of tasks and formalisation.

Strategy as a contingency factor: the strategy and structure debate
The idea of strategy as a contingent factor appeared when Chandler (1962) proposed that the choice of organisational structure depends upon the corporate strategy being pursued. Chandler's (1962) argument was based on a study of US firms over the period 1919 to 1959. Cases of General Motors, DuPont, Sears and Roebuck, Standard Oil of New Jersey showed that

structural adjustments in the organisation were implemented after strategic initiatives had been taken. These initiatives occurred in areas unrelated to the traditional lines of business: *new business development*. This study and later ones (Galbraith & Nathanson, 1979) found empirical evidence that firms first start diversification, which then leads to divisionalisation and the appearance of the multidivisional organisation. The essence of their proposition is that the organisational structure is secondary to strategy (Christensen *et al.*, 1978). Donaldson (1995) explains:

> [The]... strategy and structure fit positively affects performance and strategic changes cause a misfit and subsequently a structural change.

This idea, however, provoked a debate known as the *strategy and structure paradigm*: does structure follow strategy (as Chandler asserted), or is it the other way around: does strategy follow structure as argued by Burgelman (1983) and others (e.g. Hall & Saias, 1980)? There have been many debates in the strategy literature regarding the causal relationship between strategy and structure (most recently, Amburgey & Dacin 1994).

Burgelman (1983) refined the idea by arguing that the relationship strategy-structure depends on the part of the strategic process that is analysed. He asserted that 'structure follows strategy' generally holds true for strategic activities initiated by the current corporate strategy. However, as a firm matures (stages 3 and 4 in Chandler's growth stages), the argument of fit between structure and strategy might not be valid. Burgelman's (1983, p. 61) review of studies in the light of internal corporate venturing suggested a model of the strategic process in large complex firms. A model where current corporate strategy induces some strategic behaviour but changes in corporate strategy follow other autonomous strategic behaviour. The critical insight is that two kinds of strategic activities are suggested. Most strategic activities are *induced* by the firm's strategy, but also a number of *autonomous* activities emerge that fall outside the scope of the current strategy. This research suggests that the autonomous actions of a subsidiary, for example, can also be induced by the existing structural context. To return to the previous example, the subsidiaries of a firm, given their divisional structure, would take autonomous initiatives to expand internationally, e.g. by initiating internal corporate ventures resulting from their division structure.

While there is no shortage of literature on issues of strategy and structure, the body of empirical evidence is less impressive. Bower (1970) found support for Burgelman's argument, based on a study on the management of strategic capital investment projects. Contingency scholars generally agree that initially structure follows strategy. But it is a two-way process as argued by Hedlund and Rolander (1987).

Critiques on contingency perspective

The empirical evidence on the structural contingency has shown that the assumptions of structural adaptation are somewhat controversial. But there are validated empirical findings that performance can be partially explained by organisational contingencies (Ketchen *et al.*, 1997). Despite the contingency theory contributions, a variety of views on organisation structure have rejected the contingency theory. Over the years, contingency theory came under attack with respect to the rationality, organisational power and environment concept (Mintzberg, 1994). Astley (1985) argued that environments are quite open and receptive whatever variations are imposed on them. Van der Ven (1979) suggests that variations might be attributed to the many and different individual choices made by entrepreneurs and inventors.

Hannan and Freeman, (1977), under the label of population ecology, postulated that environmental conditions forced organisations into strategies, namely market niches. The organisation obeyed to the environment, or else it was selected out. Hannan and Freeman argued that the *power to change* rests with the environment rather than the organisation. In their opinion, the environment is complex, dynamic and hostile, dictating strategy. This population-ecology theory agrees with the contingency idea of constant environmental pressure to change, but states that the firm is not explicitly able to make choices. The contingency school assumes that firms have power to adapt to changes and survive.

The economics discipline has also contributed to the thinking on the organisation principle of fit. Transaction cost theory and agency theory (e.g. Williamson, 1975; Jensen & Meckling, 1976) are two influential contributions to organisation theory. Both approaches to organisation deal with the issue of reaching a certain *fit between conflicting interest*. In the agency theory the organisation is represented by a situation of conflict of interest between two roles known as the 'principals', e.g. company shareholders, and the 'agents', e.g. managers. On the question of how to organise, certain control structures have to be implemented to monitor the interest of both agents and principals. If not, one party will be opportunistic and strive for maximum personal benefit. Transaction cost theory holds the assumption that market failure typically occurs and hierarchical control must be introduced both to ensure effective operations and prevent managers from favouring personal goals over corporate goals.

4.4 ORGANISATION THEORY: CONFIGURATION PERSPECTIVE

Configuration approach: alignment

In contrast, the configuration approach does not take the viewpoint that circumstances have to be defined before making the combination of strategy and structure that fits. Configuration arguments are more complex than contingency arguments. It is not about fit between size, strategy, structure etc. but it is about alignment. Configurational theories differ (see table 4.2) from contingency theories because (Doty, Glick & Huber, 1993; Delery & Doty, 1996):

- they try to identify configurations that are maximally effective,
- they are usually based on ideal, and
- they assume that several unique configurations can result in maximum performance.

Table 4.2: Differences in contingency and configuration perspectives (Sources: Mintzberg, 1994; Donaldson, 1995)

	Contingency theory	Configuration theory
Message	Fit, 'it all depends on contingent factors'	Synergistic combination, alignment, 'getting it all together'
Organisation dynamics	React	Integrate
Key elements	Environmental dynamism, complexity, selection, external fit	Configuration, ideal type, external and internal fit
Environment	Broad dimensions, contingent factors like size, strategy, technology	Any
Situation	Any best fit	Any as long as it belongs to a category
Structure	Any likely bureaucratic form based on degree of functional specialisation, standardisation, and formalisation	Any as long as it is a configuration ideal type

These ideas are incorporated in both theoretical and empirical studies (Miles & Snow, 1978; Khandwalla, 1973). In the remainder of this section we will present influential work of configuration research linked to innovation.

Configurations: Burns & Stalker organisation structure typology

Burns & Stalker's (1961) study presented configurations of organisations long before configuration research emerged. They elaborated on the creative and innovative aspects of the organisation. Burns & Stalker described the opposite ends of a continuum of organisational forms as (p. 119): '... *two polar of technological and commercial chang e...*'. The first ideal type is the mechanistic organisation, supposed to be suitable for companies operating in a stable environment. At the other end is the ideal type of the organic organisation structure, supposed to be relevant to firms in dynamic and complex environments. It is seen as a more flexible structure appropriate for changing conditions. Table 4.3 summarises some configuration aspects of the organisations.

Table 4.3: Organic and mechanistic organisations (Source: Burns & Stalker, 1961)

Organic organisation	Mechanistic organisation
• individual responsibility above rules and procedures	• rigid functional departments
• cross-functional participation and dialogue	• functional specialisation
• informal internal environment	• bureaucracy
• informal channels of communication	• operational rules and procedures
• cross-functional teams operating across boundaries	• formal lines of control
• emphasis on creative interaction	• long decision lines and slow decision making
• external focus	• relatively little experimentation freedom
• non-hierarchical	• formal channels of communication
• top-down and bottom-up flows of information	• internally oriented and top-down strategy

With respect to innovation the organic structure is typically characterised by loosely coupled informal networks, making sure that routine bureaucratic aspects of an organisation do not impede the innovation process. The mechanistic organisation is characterised by such bureaucratic elements as strong formal lines of authority and hierarchical procedures. The mechanistic organisation is regarded as ineffective to innovation due its rigidity, regulated co-ordination, and hierarchy of control. Empirical studies (see: Rothwell, 1972; 1994; Cobbenhagen, 1999) have advocated the idea that innovation will seldom flourish in an organisation environment typified by mechanistic principles of organising.

Configurations: Miles & Snow strategy typology

Another straightforward contribution to the strategy & structure relationship is the well-known study by Miles & Snow (1978) on the identification of four strategy typologies. They stated that (1978, p. 7) one can conceptually associate strategy with intent and structure with action. They identified four ideal types: 'analysers', 'prospectors', 'defenders' and 'reactors'. Each type is associated with different organisational structure, processes and competencies. The configurations were derived from defining a set of distinctive functional competencies, e.g. financial management, marketing and sales, R&D, and engineering. Typical ideas were:

- Defenders, which were strong in production and applied engineering, indicating strengths in production cost efficiency.
- Prospectors, which excelled in product research and development and basic engineering, indicating their search for new market opportunities.
- Analysers which differ strongly with the industry sector, resulting in no clear standard configuration. These firms had a consistent configuration of strategy, structure and competencies that reflected the defender and prospector type.
- Reactors that did not have a logic competence pattern. They showed no consistency in strategy and structure and were outperformed by the other types. With one exception: reactors perform well in highly regulated markets.

One can observe from this study that firms pursuing different strategies do so with a different organisational configuration. The strategic differences among the ideal types suggest that organisations can pursue alternative strategies, which are accompanied by different practices and which can be equally effective in increasing performance. Many studies have found support for this typology and validated the Miles & Snow typology in various sectors (e.g. Shortell & Zajac, 1990; Miller, 1986; Snow & Hrebiniak, 1980; Miller & Friesen, 1980).

4.5 ORGANISATIONAL STRUCTURE OF INNOVATION

Cameron (1980) and others (e.g. Pascale, 1990; Stacey, 1996; Hamel & Prahalad, 1994) argued for creating creative tension in an organisation. Such a tension will keep organisation alert to new external developments, new technologies, or newly emerging competitors. In this section we will elaborate upon a number of the organisation designs for innovation. In order to use words in a consistent manner we refer to structure as: the set of formal and informal relationships between people and resources in an organisation, including the authority relationships and control systems (Donaldson, 1995).

Three influential organisation design options are presented:

- *the functional organisation*, representing functional departmentalisation and centralised authority;
- *the matrix organisation*, combining functional and product departmentalisation;
- *the team-based organisation*, using teams as the central co-ordinating device.

These options are not about adopting a best practice or a blueprint for a newly emerging organisation. Rather, they represent contributions from Galbraith (1973, 1994) and others (Nonaka, 1990; Grant, 1996). Galbraith's work inspired other contributions (Larson & Gobeli, 1988; Klimstra & Pots, 1988; Montoya-Weiss & Calantone, 1994; Cobbenhagen, 1999) on the effectiveness of these organisation designs.

Innovation in a functional organisation
A functional organisational structure represents a hierarchical and functional form of organising (Chandler, 1991). This structure is close to Burns & Stalker's (1961) mechanistic organisation structure, which is based on functional concentration and functional order of activities. Mintzberg (1979) refers to 'building an administrative hierarchy of authority:

- an *operational* part performing the basic productive tasks;
- an *administrative* part with co-ordinating tasks.

The first drawback deals with the unpredictability and difficulty to break down and integrate the separate innovation activities. Among the origins of this drawback is the appearance of functional barriers. Specialisation into functional units has the potential disadvantage of creating functional barriers between departments, for example between R&D, production and marketing. These barriers can result because of (Griffin & Hauser, 1996):

- different task priorities;
- unclear responsibilities between functions;
- control systems that evaluate activities differently;
- dual rewarding and career development schedules;
- lack of top management support.

Several authors (e.g. Clark & Wheelwright, 1993; Nonaka, 1990) have argued for the need for cross-functional co-operation as a means for closer integration between functions. Two organisational mechanisms for integration are (Grant, 1996) direction and routine. Direction relates to the presence of formal procedures (programming), and the distribution of authority, control of tasks and responsibilities (Grant, 1996). It can take many forms and is often embedded in the management systems of the organisation. Procedures are explicit and refer to structure (Giddens, 1979). Such formal procedures include project planning techniques, investment decision and budget control methods.

Routines are implicit and refer to action (Giddens, 1979; Nonaka, 1990). They might be compared with the 'built-in patterns of action' and the way people operate in an organisation. Routines are embedded in the people working in the organisation (Nelson & Winter, 1982). Often the company culture is associated with routines (Schein, 1985). The relevant question here is whether routines can contribute to better integration. If the integration of separate functional tasks can become a built-in routine, people will share their knowledge automatically. As an example, if market researchers share their first findings on consumer trends and share them at an early stage with R&D people, this will contribute to integration of such tasks as concept development. Routines are 'path-dependent'; similar to the ideas developed in the resource-based view of the firm (Teece *et al.*, 1997). Or as Nonaka and Takeuchi (1996) state: routines, just as capabilities, are socialised in the organisation.

Innovation in a matrix organisation

The matrix design has received considerable attention in the literature since the major contributions by Galbraith (1973). Because the matrix organisation is a complex and recent response to both the needs for specialisation and co-ordination, its effectiveness is difficult to assess. The matrix form fundamentally breaks with the basic principle (Fayol, 1949) of organising: unity of control. Research findings (Cobbenhagen, 1999; Larson & Gobeli, 1988) have shown that a matrix-like organisation can be more effective in enhancing project integration and speed (Larson & Gobeli, 1985) and meeting client needs. In this study we expect ICT firms to be progressive in the use of matrix-like structures. But let us look at a theoretical explanation first.

Galbraith (1973) described the matrix organisation as a structure model with dual lines of authority. The functional line organisation continues to exist, but in addition an overlay of a project line is introduced, which would co-ordinate the activities of the departments. This structure is characterised by functional lines of authority that are deliberately combined with the existence of formal project teams. This construction would ensure proper communication and co-operation across functions in an innovation project. Studies (Cobbenhagen, 1999; Barczak, 1995; Larson & Gobelli, 1985, 1988; Klimstra & Potts, 1988) have shown that matrix approaches can work well as a temporary or permanent design.

Disadvantages mainly result form the inherent conflict of interest of the functional line and the project line (Klimstra & Potts, 1988). This leads to the appearance of dual reporting lines causing decision-making problems. Duality in communication, co-operation, and reporting might be considered as the potential organisational constraint. It may result in a lack of clarity on work expectations and work performance between departments and with the project line. Also, conflicts with regard to resource availability and allocation can crop up due to different organisational demands. These conflicting situations can induce the emergence of politics by project and functional managers to reach certain goals. Furthermore inefficiency may arise as a result of 'non-co-ordinated' duplication of activities.

Innovation in a project team organisation

Compared to the matrix organisation the project team structure is extended along the project axis (Larson & Gobeli, 1988). Characteristic of this form is a project leader who is put in charge of a project team composed of a core group of people from all relevant functional disciplines. They are allocated to the project on a full-time basis. Functional managers do not have any formal involvement. Cross-functional teams are assumed to be critical to innovation performance (e.g. Clark & Fujimoto, 1991; Dougherty, 1992). Table 4.4 briefly summarises the differences in design options.

New insights: Lateral organisations
The design of lateral linkages (Den Hertog & Huizenga, 2000) in an organisation can help to surpass organisational boundaries. Project team structures are suggested (e.g. Pierce & Delbecq, 1977; Larson & Gobeli, 1988) to be effective in designing such cross-functional linkages. The suggestion of lateral links, without burdening the hierarchy, is a rich concept further developed by Galbraith (1973, 1994). Galbraith refers the organisation designed with cross-functional linkages as the lateral organisation. Lateral relations cut across functional lines of authority. They allow for decentralisation of decision making to the lowest possible level of

Table 4.4: Summary of organisation design options

	Functional organisation	**Matrix organisation**	**Project team**
Focus	Functional specialisation	Functional lines and formal project lines	Multidisciplinary teams
Organisation flexibility	Weak	Medium	High
Resource efficiency	Medium, reduces duplication of scarce resources	High, reservoirs of specialists	Medium, project-driven
Interdepartmental relationships	Multiple, simple	Multiple, complex	Multiple, within team
Line of reporting	Single line, functional	Dual line	Single line, project
Manager	Functional manager	Lightweight project manager and functional manager	Heavyweight project manager
Cross-functional integration	Weak	Moderate	Strong
Customer focus	Weak	Moderate	Strong
Human resource management emphasis	Functional	Cross-functional	Cross-functional

Various sources: Cobbenhagen, 1999; McCann & Galbraith, 1981; Roussel *et al.*, 1991

action. The result of such structures is an increase in the capacity of the organisation to absorb and process information and make decisions rapidly. Next, we will describe the ideas on project teams and multidisciplinary teams. This team concept is supported by research on the organisation of innovation (Leonard-Barton, 1995; Dougherty & Hardy, 1996; Cooper, 1994; Pisano & Wheelwright, 1995; Clark & Fujimoto, 1991; Souder, 1987).

Heavyweight project teams

Several authors (e.g. Clark & Wheelwright, 1992; Clark & Fujimoto, 1991) conducted empirical research in the automobile industry, which is supportive of the use of multidisciplinary teams. Clark & Wheelwright (1992) introduced the concept of heavyweight project teams, which is typical of large-scale innovation projects. Heavyweight project teams are multidisciplinary teams with a heavyweight project leader who acts as a powerful linking pin and co-ordinates the activities of the product development team, but also works with senior management to create an overarching product concept. In this concept the senior management can exercise subtle control through such project leaders, who manage the multidisciplinary team in the context of a product vision. This type of team structure offers improved communication, stronger identification with and commitment to a project target, and a focus on cross-functional problem solving.

Multidisciplinary team design options

Multidisciplinary team concepts appear in various forms, such as *rugby teams* (Takeuchi & Nonaka, 1986), *skunkworks* (Quinn, 1985), or *venture teams* (Bart, 1992). Tom Peters (1988) defined skunkwork as highly innovative, fast-moving, and slightly eccentric activities operating at the edges of the corporate work. The origins of skunkworks go back to a development team at Lockhead's Design and Development Centre. These cross-functional teams operate for a certain period on a new product concept and are fully detached from the regular organisation and daily practice. They often have a separate location and work under tight deadlines. In contrast, venture teams operate in an environment with high-risk endeavours. It is uncertain if this specific team approach is favourable above other team structures, as they are infrequently used for new product development (Bart, 1992). Each of these multidisciplinary team concepts resembles what Burns & Stalker (1961) refer to as the organic structure of organisations. When we regard the advantages of team designs and record the variations in team concepts it is likely that similar design options will appear in the ICT sector.

4.6 ORGANISATIONAL CULTURE

Successful innovation involves more than designing structures and systems. Investments in the quality of innovation are also affected by the organisational culture of a company. A dominant view in studies on organisational culture is that the development of a culture is based on the shared experiences of organisational members (Schein, 1985). Culture then creates a shared belief between organisational members (Pfeffer, 1981). Influential work by Schein (1985) argues that culture contributes to enhancing (but at some stage inhibiting) the internal integration and the adaptation of the organisation to change. Organisational culture is like a 'glue' for aligning internal processes and adaptation to changing market conditions (Kets de Vries, 1991; and Scott Morgan, 1994).

Different descriptions and definitions of culture have been developed. Schein (1985) asserts that the analysis of culture can take place at three subsequent levels:

- The first culture level of an organisation is maintained through artefacts and creations. These are more explicit like managerial policies, educational programs, and investment principles (Leonard-Barton, 1995). They result from experience accumulation but are not well documented. They identify the 'gut feelings' and 'rules of thumb' that direct actions and decisions.
- The second level incorporates the values of a company, which are espoused, overt and often subject to debate. They are more deeply rooted in the organisation's systems, behaviour and processes.
- The third level involves basic characteristics of human nature, communication and relationships. The values that are generic are often cited as the corporate values of a company (Leonard-Barton, 1995). They form a set of strongly related assumptions that are shared by all company employees. A strong culture might indicate that values are anchored in the organisation and difficult to change as they are widely shared by employees. Studies (e.g. Scott Morgan, 1994) have documented that the cultural barriers ('unwritten rules of the game') can be strong and can impede renewal and innovation.

Culture and innovation
There are many research studies on the cultural dimensions of an organisation and their effect on behaviour and performance (Chatman &

Jehn, 1994; O'Reilly *et al.*, 1991; Schein, 1985). In the innovation literature, culture is often regarded as the final organisational ingredient essential to make innovation happen (Cooper, 1996). Some specific studies describe 'the way an organisation operates' (e.g. Cobbenhagen, 1999; Moss Kanter, 1983) and have recorded the importance of culture for innovation success. They have identified that the values and norms have a profound effect on innovation and building up knowledge and capabilities (Leonard-Barton, 1995). Organisational characteristics such as openness, respect, teamwork, external openness, and commitment to learning are assumed to be supportive for innovation success (Leonard-Barton, 1995). Bolwijn & Kumpe (1992) also assert that technological innovation is strongly related to culture. They state that (1992, p. 75):

> '...Innovation is a dominant competition criterion, requiring a company where behaviour is minorly formalised [...] There is an atmosphere of openness. Thinking in alternatives is strongly developed. Across the organisation there is much information exchange, [...] much communication is informal...'.

This leads us to say that it is likely that firms that are successful in innovation have an organisation culture that differentiates on such dimensions as informal, openness, extrovert and low formalised rules. We will investigate whether ICT firms have a similar profile.

4.7 ORGANISATION: THE HUMAN CAPITAL FACTOR

Recently scholars have argued that knowledge is an important production factor in innovation (e.g. Den Hertog & Huizenga, 1997, 2000; Nonaka & Takeuchi, 1996; Drucker, 1993). Part of the knowledge is tacit knowledge (Polanyi, 1966), which resides in the human resources in an organisation. An innovation process is driven by the development and sharing of both this explicit and tacit knowledge. Explicit attention for the human resources in an innovation process can contribute to improving the development and sharing of tacit knowledge (e.g. Leonard-Barton, 1995; Nonaka & Takeuchi, 1996). This notion suggests a policy role for human resources management in strengthening the ability to learn and to innovate in a company. We expect these topics to be of great importance to the ICT sector as instruments to intervene in knowledge development processes.

Human resource management: recent contributions

In the literature contributions (Huselid *et al.*, 1997; Schuler & Jackson, 1996) there is a call for a strategic orientation to human resources, which entails the design and implementation of a consistent set of instruments and policies. Becker & Gerhart (1996) and Pfeffer (1994) suggest that human resource policies should be sufficiently rare and inimitable to create a sustainable competitive advantage. This extends the 'traditional' idea of a technical, cost-oriented personnel policy (Huselid & Becker, 1996; MacDuffie, 1995; Van Sluijs & Den Hertog, 1993). In general, the HRM philosophy proposes that human resources are a valuable knowledge source that is difficult to imitate (e.g Den Hertog & Huizenga, 1997; 2000; Becker & Gerhardt, 1996; Pfeffer, 1994; Barney, 1991), and that HRM involves the methods, policies and practices that allocate knowledge to the most productive use (Drucker, 1993). The instrument panel of the HRM field (Den Hertog & Huizenga, 1997; 2000) can be used to steer the development of knowledge in the organisation and improve innovation performance (Huselid *et al.*, 1997). What main practices are relevant to ensure a cross-functional focus?

Cross-functional mobility

There exists strong evidence (Griffin & Hauser, 1996; Song & Perry, 1996, 1999; Souder, 1987) that shared commitment and co-operation between marketing and R&D enhances success, for example, in setting new product goals, identifying opportunities for product improvement, resolving engineering problems and customer needs. Stimulating the cross-functional mobility might be achieved by putting people in various jobs in various functions or projects. In the R&D philosophy this has been documented as a *dual career ladder* (Allen & Katz, 1986). The idea of the dual career ladder is that R&D people have the opportunity to develop themselves along a management trajectory or a research trajectory. This dual ladder can be mono-disciplinary in nature, if one specialises within the R&D context. From the viewpoint of cross-functional mobility, this ladder might need a third career stream: a 'pyramid ladder'. A pyramid career would allow for a career development along three dimensions: the management level, deeper functional skills and broader cross-functional jobs.

Career development & planning

Case studies record (e.g. Den Hertog & Van Sluijs, 1995) that people become personally responsible for their career growth. In the past, firms provided lifetime employment and a career pattern. Individuals recognise that their employability increases if they acquire skills that make them more flexible for job switching and rotation. Career development plans can make explicit which path an individual has to follow to achieve future job positions and become multidisciplinary.

Recruitment & selection

Typical of a 'people business' is the recruitment, selection and coaching element. We expect ICT companies to be keen on these subjects as well. One way to acquire knowledge is the recruitment of people who have the necessary skills to perform a job. Recruitment and selection can take many forms, like formal selection and interview procedures, internal job pools or external networking. The selection policies are based on the principle of 'matching job requirements with the employee's knowledge and skills'. Aligning these practices with the innovation could mean that the recruitment practice focuses on people who are apt to learning.

Coaching & rewarding

Coaching of human resources is often oriented at supporting current job fulfilment and less on knowledge development. Coaching as a steering instrument can be supportive to innovation. It is used to identify differences between current and future requirements of knowledge and the skills within a person's individual profile. Steering on individual development plans aims to upgrade the knowledge level. Complementary to that are reward systems, known as one of the most powerful motivators in the organisation. Rewarding can be a direct instrument for evaluating an individual's contributions and progress.

Education & training

This topic is a well-understood and obvious HRM instrument in the knowledge-intensive business sectors. We expect it to be a self-evident subject in the ICT sector. If education focuses on deepening of functional skills ('more of the same') and not on broadening skills this might risk the development of new knowledge (Mohrman, 1992). To prevent this from happening education and training policies can be used as a steering instrument for cross-functional knowledge sharing and development. This can take many forms like formal training and development programs. Furthermore, there is a risk that knowledge becomes obsolete over time. People forget, unlearn, are stuck in routines, which hampers innovation. Even more, due to the intensified market competition and the ongoing urge for value and productivity increases, knowledge becomes obsolete quickly.

PART II:

Research Design & Methodology

5. The Case Survey Research Method

The use of multiple data collection and analysis approaches facilitates exploratory research. Given our research interest in studying change, we constructed a new hybrid methodology. The questions raised in this study require a broad perspective of the management, strategy and organization of ICT companies and their respective business environment. Analysis of innovation and change in the ICT sector implies that both software and service innovation are included in the study, primarily because of the close interdependence of products, services, processes and technology. A major change in a software or hardware product influences other producers and service providers along the ICT value chain. Insight into the variety of business of ICT companies is thus necessary to understand innovation in the ICT context (Abernathy & Utterback, 1978). These circumstances require a research design that takes account of these issues.

5.1 INTRODUCTION

A research strategy, or design, is a process of collecting, analysing, and interpreting observations (Yin, 1989). It is regarded as a 'logical model of proof'. Within social sciences different research strategies prevail to analyse complex situations. The most dominant strategies are the experiment, the survey, and the case study. The experiment is a specific form of research as it involves researcher intervention. For this study it was not an appropriate strategy. The attraction of the research strategy largely depends upon the research questions asked and the purpose of the research, e.g. being explanatory, descriptive or exploratory. The choice of strategy is determined by the initial questions asked. Research is essentially a process consisting of a range of process phases. For each phase one can pursue different research strategies that have their own specific value. These value-driven strategies are based on answering 'who', 'where', 'when', 'what', 'why', or 'how' questions.

The dual methodology used in our research strategy combined the following elements:

1. case studies based on in-company interviews with four persons;

2.   an in-company large scale survey, consisting of over 500 variables.

This combination can be referred to as a case survey.

The Case Study
Case studies are a preferred research type if the three following conditions apply (Yin, 1989):

- if how and why questions are being asked;
- if the investigator has little or no control over events;
- if the focus is on contemporary phenomenon within a real life context.

Most case studies are of a retrospective nature because they describe past events. Reports, documents and business plans provide insight into past actions and strategic and organisational choices that have been made. Case comparison and analysis reveals a number of underlying factors and patterns. This kind of analysis can be called 'detective work' and is designed to reveal patterns and consistencies between cases (Mintzberg, 1979). A traditional comment on case study research is concerned with its lack of comparison, making generalisations difficult. However, there is an essential difference between qualitative and quantitative research in this respect. The basic premise of quantitative research is that statistical inferences can be drawn that provide generalisations going beyond the sample. Qualitative research does not aim at statistical generalisation but at theoretical generalisation (Yin, 1989). Yin argues that the case is an experiment to falsify some theoretical proposition. Yet this implies that the case selection is not a random but a carefully selected process for theoretical generalisation. One case study, however, cannot provide the entire setting of a phenomenon (Yin, 1989). Multiple case design is a method to increase the generalisability, or external validity, of the outcomes. Contrary to generalising from increasing the number of cases and degrees of freedom (statistical generalisation) in quantitative research, qualitative research contains a 'replication logic' (Yin, 1989). Yin (1989, p. 53) proposes two different logic's of replication:

- theoretical replication: being i.e. the careful selection of cases to provide results for rejecting previous reasoning;
- literal replication, aimed at: predicting similar results.

Use of Case Study Material

Pettigrew (1990) recommended that when research is characterised as exploratory or theory building, it is particularly important to begin with an 'extreme case', i.e., a setting where the phenomena of interest are frequently occurring and readily transparent. Because this study of innovation attempts to explore the validity of key success factors and explore new constructs to contribute to resource-based theory, the ICT industry provides an ideal context for this research. A large number of companies visited in the ICT sector have argued in their external communication and during the interviews that the 'unique' or 'idiosyncratic' nature of their firm distinguishes them from their competitors. They implied that their business strategy, organisational design and human resource policy differed from others. This would indicate heterogeneity of firms in the ICT industry. The resource-based approach (Barney, 1991) assumes that this idiosyncrasy of tangible and intangible factors upon which strategy is based, can explain performance differences. The research field can be strengthened by empirical investigations of the resource-based view that incorporate the constructs accompanying this approach.

The cases were primarily used for three purposes:

- Cases have been used for illustrative purposes. Case descriptions have been entered to confirm the statistical findings. This can both support the statistical results and clarify its context. This reduces the risk of drawing wrong conclusions from quantitative data and supports the generalisation of findings. We are not interested in the statistical generalisation to a population. We use a meaningful sample to understand the relations and the process and context behind innovation.
- Case descriptions have been admitted for providing alternative explanations. Whenever statistical outcomes could not be explained, the case material could explain the unique character of an outcome or give an alternative reasoning. Whenever possible we listed case material to support alternatives.
- Cases were carefully used to test specific assumptions that could not be explored with quantitative technique and collection methods. Although this does not result in statistical generalisations it can support sensitising concepts and add to analytical generalisation.

In this way we have tried to triangulate the data. There exists a fourth purpose for using cases, namely the possibility of a comparative case study research. This entails comparison of the case data along the guidelines of Grounded Theory (Glaser & Strauss, 1967; Strauss & Corbin, 1990). However, this process of structuring, coding and comparing case data fell outside the main goals and framework of this study.

The Survey
A survey is a collection of research methods that are used to develop statistics within a population or a representative sample (Fowler, 1988). A survey is a research method specially designed to answer 'what', 'where', and 'when' questions. Its advantage over case studies is the ability to calculate whereby systematic measurement yields a data matrix that can be analysed to reveal patterns. In the data matrix relations can be found between the variables. There are three preconditions for the existence of causal relations (Den Hertog & Van Sluijs, 1995):

- covariance;
- exclusion of alternative explanations;
- a one-directional relation between cause and effect.

The most important drawback of the survey is that it is incapable of providing a meaningful interpretation of why things happen. Although surveys are not suitable for these research questions, it does happen that studies try to understand complex processes by designing a survey. The second drawback is the lack of excluding alternative explanations. A check on the non-response is one way to decrease the set of explanations. Another problem with surveys is its time dependency. Leonard-Barton (1990) argued that measures are sensitive to the point in time at which they are administered. Survey studies often leave the researcher with more questions about the 'how' and 'why' of a phenomenon. In that case, qualitative data can enrich the understanding of the innovation process.

In this study we conducted a combination of case study and survey. This research strategy is called a *dual methodology*. The sample of the study will be large enough to draw statistical conclusions and meanwhile interpret the results within the context of the study.

5.2 DUAL METHODOLOGIES

What is the status of case study research in innovation research? Empirical research related to dynamic capabilities and competencies perspectives is grounded on descriptive case studies from successful firms (Collis &

Montgomery, 1995; Hamel & Prahalad, 1994; Nonaka, 1991). These case studies have provided significant insights into the nature of competence-based competition. Recent contributions by Miyazaki (1995), McGrawth (1995), Iansiti & Clark (1994) and Henderson & Cockburn (1996) have enriched the field with survey data. These studies provide us with the opportunity to test the rigourness and sensitivity of the concepts and measures over time (Leonard-Barton, 1990).

Research approaches to new topic areas often rest upon novel combinative research methodologies like 'blended' and 'dual' methodologies (Den Hertog *et al.*, 2000). Theory building research typically combines multiple data collection methods (Eisenhardt, 1989). Eisenhardt and Bourgeois (1988) combine surveys with qualitative evidence from interviews and observation to describe strategic decision making in high-velocity markets such as the computer industry. Womack *et al.* (1990) consolidated data from different sources in their MIT study. Work by Jick (in: Den Hertog *et al.*, 2000) promotes the use of multiple sources of evidence by advocating triangulation of data types, providing reliable constructs and better hypotheses. Mintzberg (1979, p. 587) described the importance of triangulation by stating:

> For while systematic data create the foundation for our theories, it is the anecdotal data that enable us to do the building. Theory building seems to require rich description, the richness that comes from anecdote. We uncover all kinds of relationships in our hard data, but it is only using this soft data that we are able to explain them.

The case survey
This strategy has appeared in several forms (Den Hertog & Van Sluijs, 1995). The case is regarded as the unit of analysis and attention is paid to the context of the change processes that are analysed. Four categories of case surveys can be identified (figure 5.1):

1. A survey within one case: Leonard-Barton (1990) performed a survey administered to several employees within one company (case study).
2. Case sampling and a survey from codified data: Larsson (1993) proposed a design by drawing a sample of cases from a larger number. For each case qualitative information was coded into quantitative data allowing for statistical analysis. This procedure

uses a strict coding scheme and independent interpreters to transform data.

3. A large survey with one individual case: the unit of analysis is an exemplary case that can be a project, programme or firm. For this case some contextual data is acquired (Van de Ven & Poole, 1989).

4. A smaller survey with more cases: each case is subject to a complete survey. Cases consist of a complete set of interviews, documents and case descriptions. A large survey data set is available for each case (Den Hertog & Huizenga, 1997).

Figure 5.1: Case survey methodologies

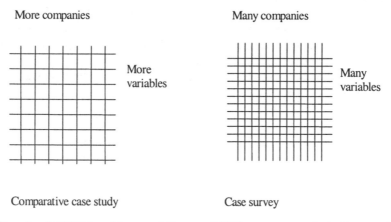

Eisenhardt (1989) and Leonard-Barton (1990) argue strongly in favour of the use of dual methods. Combining methods creates synergistic advantages and can indicate salient relationships that would not have been detected if one method had been used. At the same time it can support findings or prevent false interpretations. Qualitative data contribute to the understanding of the quantitative relationships and strengthening them. Combining research methods has advantages and disadvantages, which will be discussed below. Table 5.1 provides a summary of comparison between the methodologies used.

Leonard-Barton (1990) provided two sets of advantages for combining different methodologies in a multiple cases and longitudinal case study on technology transfer:

- specific strengths in the data-gathering process for each method compensate for particular lacks or weaknesses in the other, and
- complementary approaches in each method enhance validity because of synergy obtained by combining methods (p. 255).

5.3 (DIS)ADVANTAGES OF A DUAL METHODOLOGY

The strength of a case survey is that one can generate covariance to analyse the possibility of causal relations. Because of the context of the case data one is able to generate alternative explanations for statistical relations identified.

Biases
The advantages of a case study are its direct observation, systematic interviewing and full variety of sources of evidence. One disadvantage of case study research is the danger of biased views influencing the direction of the findings. However, this bias can also enter in the design of a survey (Yin, 1989). The critique of the low generalisability of cases is safeguarded through designing and administering a large-scale survey.

Context
Case studies are also typical instruments for placing a phenomenon in a historical context (Den Hertog & Van Sluijs, 1995), allowing for the description of changes, processes, organisational choices. The advantage of multiple cases is that concepts can be better defined and general patterns can be recognised in different environments (Eisenhardt, 1989).

Dyer & Wilkins (1991) point out that multi-case research remains too general and fails to go into the deeper meaning of events. The collection of detailed data at multiple sources and multiple organisational levels can compensate for this drawback. The coherence between successive events, e.g. strategic and organisational choices with respect to innovation, is better addressed by using cases rather than by proving correlation relationships (Mohr, 1982). The advantage of case research is that one takes into account the type of firms and the markets they serve. In the study we tracked down the strategic development and path dependency of the firms (Huizenga, 2002a). In the interviews and the survey, we have used time-dimensions to track the origins of the firms and the strategic directions that have been chosen.

Alternative explanations
One important element of case study in relation to a survey research is concerned with explanation building. Case study research can consider plausible *alternative explanations*. Surveys cannot; if they did they would endanger the internal validity and hence one of the preconditions for causal relations. The combination of both methods provides the opportunity of reviewing alternative explanations when no clear relationship between variables is detected in the first instance.

Advantages of dual research methods are based upon the combinative strengths of both methods. However, there are some disadvantages as well, which are concerned with the design and the data collection. Leonard-Barton (1990) observed co-ordination problems and information overload as being disadvantages inherent in the design. Operational problems encountered in their study were based on difficulties in data structuring, in defining the unit of analysis and in finding the rational choice for the selection of cases. Much time and effort are required to co-ordinate the collection of data.

Unit of analysis
To fully understand the complex interactions underlying innovation the research methodology incorporated an *embedded design* (Yin, 1989). An embedded design denotes several levels of analysis. This provides the researchers with multiple perspectives to explain factors underlying the process of innovation. In this study the firm level and project level is examined. We used multiple sources of information and (functional) perspectives to establish a chain of logical reasoning (Den Hertog & Van Sluijs, 1995).

Disadvantages of case survey
The danger of conducting only a survey or case study is that is might lead to irrelevant or unreliable results. The advantage of a case survey is that different research strategies support the outcomes of a study. However, there are also a number of practical disadvantages. First of all, the interpretation and continuous interaction between qualitative and quantitative data calls for a carefully structured method, requiring strong research skills with regard to analysing and combining data. Second, case descriptions are not as extensive as in the case of a single or multiple case studies. The in-depth knowledge gained in a case study is more than that in a case survey. The case description of a case survey tries to construct the 'Gestalt'. Third, a case survey does not always yield statistically significant

results. This is mainly due to the small sample size. The size of the sample is a choice between 'knowing a lot from a select sample' or 'knowing little from a large number of units in a sample'.

Table 5.1: Overview of the methodologies (adapted and adjusted from Leonard-Barton, 1990).

Research phase	Multiple cases	Survey
Data collection:		
Efficiency	Relatively high, danger of irrelevant data gathering	Focused data collection by means of precoded questionnaire
Objectivity	Danger that information is subject to biases from respondent and researcher biases	• no interference by researcher • no control over correct interpretation of precoded questions (halo-effect) • multiple sources
Pattern recognition	Pattern matching across cases	Statistical generalisation
Quality of design:		
Internal validity	Cause and effect relations difficult to recognise	Covariance and correlation analysis
External validity	Theoretical generalisation	Statistical generalisability
Construct validity	Validate constructs across events	Operationalisation of constructs through pre-testing questionnaire

In sum, dual methodologies are combined research strategies with the intention of grasping both 'what', 'why', and 'how' questions within one study. Both survey data and contextual information are collected to generate a complete picture of a phenomenon. In the next paragraph we will go into detail on the design of the case survey.

5.4 RESEARCH DESIGN

Definitions
In order to ensure a consistent meaning of innovation throughout the research study we have defined innovation as follows (OECD, 1994; Tushman & Moore, 1992):

- Product and service innovation: important improvements of existing products and services and the development and commercialisation of new products and services.
- Process innovation: important changes and improvements of existing processes and the development and implementation of new processes.

Basically, we will address in this study all of the above-mentioned aspects, e.g. the focus, the source and the impact of innovation. All things that are new to the company are considered to be relevant to innovation in the ICT sector because the sector is knowledge-intensive and facing continuous and rapid technological progress. However, we expect the managers in the study to particularly address such subjects as the creation of competencies, the radical role of technological change versus the presence of incremental innovations and the benefits of new or improved products, services and internal processes to their customers.

Cooper (1996) suggested two definitions of newness:
1. New to the company, in the sense that the firm has never made or sold this type of product, service or process before, but other firms might have.
2. New to the market or 'innovative': the product is the first of its kind on the market.

Successful innovation was defined as *the economic exploitation of innovation*. 'Turning good ideas into good currency', as Van de Ven (1986) suggested. This enabled us to measure success in terms of achieving cost, profit or turnover objectives, meeting customer needs or technological success. We recognised that performance depends to a large extent on the types of new products, services and processes. The impact of innovation efforts is not measured at one moment in time. A broader concept would have to take the life cycle of innovations into account. Furthermore, by including multiple measures both the development and management of new products, services and processes are taken into account (Barclay, 1992). To summarise, the idea is to address the success factors based on a multidimensional framework. Although there are no ready-made procedures for this method several innovation studies provide examples on the use of this method of success measurement.

In order to ensure a valid replication of the study we also addressed the first definition. In addition, we based performance on common practice in the product development literature (Griffin & Page, 1996; Brown & Eisenhardt,

1995). A part of the success variables we selected belong to the category of managerial perceptions of innovation aspects. These variables were gauged using a 1-5 Likert-type scale. These variables were combined with a second category of financial data, which were directly measured as percentages or absolute values. All measures were developed to address firm level innovation, i.e. performance measurement was not restricted to a single product, project or programme level, but included the total scope of a firm's product and process innovation efforts

The Survey Questionnaire

For the survey a questionnaire was used. A questionnaire is a highly structured data collection technique where respondents are asked a same set of questions (De Vaus, 1991). This study uses an existing and validated survey instrument on innovation (Huizenga, 2001; Cobbenhagen, 1999). The questionnaire was evaluated and customised to make it appropriate for the study of innovation in the ICT sector. The process of designing, testing and administering the survey entailed the following steps.

Scaling

For most variables a Likert-type scaling was used to measure attitudes (Likert, 1932). The questionnaire includes over 500 variables. Respondents were asked to score on each questionnaire item using a 1-5 Likert scale, where 5 indicates a larger value of the variable. Financial data were gathered and for some variables percentages were used for the scale. Some questionnaire items were presented to one respondent and others were presented to all the respondents in one company. We interviewed four persons in each company. A changing team of two researchers conducted interviews with the marketing manager, R&D manager, and managing director in all companies. A fourth person, a financial manager, senior project leader or personnel officer, was consulted for company background information and financial data. . Sometimes initial in-company interviews were held to ensure that the proper unit of analysis and correct sources of information were used.

We expected to find differences in the responses from the people interviewed, e.g. the cultural variables, the degree of cross-functional co-operation and the external co-operation with customers and suppliers. To include a path dependency element in the questionnaire, a time frame was used for some variables. Similarly, the variables on strategic priorities were measured for two periods, 1988 to 1992 and 1992 to 1996.

Use of self-rating scales
One might argue that the use of self-rating scales affects the reliability of information. The use of self-rating scales for innovative success has been used extensively (Cooper & Kleinschmidt, 1995). However, concern has been expressed that these scale types are reliable and unbiased. The use of self-reporting scales as performance measures has been criticised for its subjectivity and danger of overestimation by respondents (Chenhall & Brown, 1988). To prevent the collection of unreliable data, multiple sources have been used. Furthermore, items have been developed that describe the organisational routines in a firm. Using the questionnaire we collected data on facts that were not subject to subjective interpretations my managers.

Brownell & Dunk (1991) argue that the nature of any bias by self-reporting measures has never been described or shown to impair the inferences drawn from the data. Venkatraman & Ramanujam (1987, p. 10) support this observation by concluding that: '... neither type of measure is intrinsically superior to the other 'in terms of consistently providing valid and reliable' measures of performance ...'. There is no convincing evidence that objective measures, e.g. financial data, are either more reliable or valid in studies (Brownell & Dunk, 1991).

Several studies on competencies use self-reporting scales to grasp the tacit concept of competencies. We asked the respondents to report on objective organisational and strategic items. Spector argued that the distortion of information under this condition is more limited than when subjective characteristics are being rated (McGrawth *et al.*, 1995, p. 266). Crampton & Wagner (1994), on the basis of a meta-analysis of 42,934 correlations published in 581 articles, also argue that self-report measures are not less reliable than objective measures. In their study on team performance in new product development, Ancona & Caldwell (1992) use subjective ratings for measuring competencies. They argue that subjective performance ratings are most often used to make budget and promotion decisions; they are related to final performance evaluation, and objective results are often the resultant of subjective ratings.

Non-response survey
The use of a check on non-response (De Vaus, 1991) increases the reliability of a study. One must assume that the non-responding firms present a different set of characters. To check if the non-responders were different, we conducted a check on the non-responding firms. From the initial results we selected the most statistical significant variables. A two-

page fax-survey containing these variables was designed and sent to the non-participating firms. The results showed no differences between the firms that belong to the sample and the non-responders. This increases the reliability and generalisability of the conclusions.

Data collection
During the pre-collection phase several steps were taken to acquire an overview of the industry. Open interviews were conducted with 15 experts in the sector, representing both the business perspective and the academic discipline: software clients, researchers, industry analysts, ICT consultants and the management board of the sector organisation. This research phase aimed at testing the questionnaire that had been developed and gaining information on the industry structure. The information, consisting of market figures and growth perspectives, was stored in a database. and consulted throughout the research period.

We conducted the research between 1995 and 2001. In-depth semi-structured interviews were held with 125 managers in 32 ICT companies, following a pre-developed interview protocol. Managers were asked to tell their story with respect to innovation in their organisation and the success of their innovative efforts. This way the researchers deliberately tried to create an interview atmosphere where the managers would not feel any constraints because of predefined questions that were unsuitable or irrelevant. During the interview the researchers posed more detailed and specific questions on such issues as business strategy, organisational design, markets, technology and innovation processes. All predefined questions were presented in the format of a benchmark questionnaire. Initial questions asked were concerned with the company development (paths), the current position, and the future direction. These interview topics are very similar to Teece *et al.*'s (1997) three dimensions of dynamic capabilities, e.g. position, paths and processes. The intention of these interviews was to hear the company story from objective sources. The managers that were interviewed had been selected by means of a profile that had been sent to the firms in advance. Additionally, we selected relevant company documents, e.g. business plans, project management procedures, annual reports and company reports. Secondary data sources were consulted to obtain information from the companies, e.g. IDC, Gartner, EITO. Also a database was developed containing a collection of web sites, professional journals, newspaper articles, and research reports.

5.5 ICT STUDY SAMPLE

The aim of this study was to present a broad picture of innovation in the ICT industry. Therefore we opted for a meaningful sample, rather than a random sample for statistical generalisations. The research concentrated on all activities across this industry. The emphasis was to be more on the complete range of ICT activities than on the statistical representativeness. Another argument for adopting a cross-industry perspective is the existence of 'mixed companies'. These firms perform the complete range of the ICT chain, e.g. software development, implementation and system maintenance. This strengthened our belief that the sample should include the whole range of ICT activities, instead of focusing on one ICT activity alone. In order to obtain a complete image of innovation in the ICT sector we intended to analyse innovation in such activities as software-package development, software tool development, system maintenance and control, and posting ICT personnel.

At the same time we also wanted to include small and medium-sized firms as well as large firms in the sample, once again with the purpose of obtaining, a broad picture. Furthermore we selected firms that conduct all business functions that are relevant to innovation. This implied that firms with large R&D centres located outside The Netherlands (e.g. Silicon Valley) were excluded from the sample. These firms performed their own technology development. All firms should perform their own marketing, production and R&D activities. We excluded firms with less than 20 employees from the sample, in order to ensure the presence of all business functions. As a consequence hardware companies fell out of the range of the study because there is only one hardware firm in the Netherlands that performs its own R&D. These initial requirements have been composed in order to control for the autonomy of the firm (Child, 1984).

Sample selection procedure
A database from the sector organisation FENIT (ICT Netherlands) has been the starting point for the sample selection. This organisation represents 180 ICT firms in Europe, and represents 90% of the Dutch ICT market in terms of turnover. Before selecting the firms we defined three classes of ICT companies:

- *Software developers*, whose main activities are basic research, applied research, development and distribution of standard application and system software packages and development software (case tools).

- *System integrators*, whose main activities are services for system and network integration, and the development of customised software, including consultancy and software-maintenance and training.
- *System and network control and maintenance*, being services firms that provide outsourcing and outsourcing services for systems and networks.

Sample selection took place in two parallel steps. First, all of the 180 FENIT members were invited to participate. On the basis of this invitation, 15 firms agreed to participate right away. At the same time we took a random sample of 70 firms from the population of 180 FENIT companies. These companies were contacted by both email and telephone. In parallel, we invited 80 firms from a database of 2,749 companies, of which 2,500 appeared to be appropriate. This database contained a lot of smaller ICT companies, many 'mixed companies', as well as companies that perform irrelevant activities. This resulted in 32 firms that wanted to participate in the study.

Sample limitations
The nature of the study implies that it is limited by the traditional pitfalls that exploratory research entails. Work remains to be done to generalise the conclusions beyond this sample, as they need to be tested in a larger sample. Because of the small sample size and the non-randomness of the sample units one needs to be careful with respect to the interpretations that have been made. A more comprehensive and larger research analysis can provide important insights for the development of competence-based perspectives of innovation. However, the richness of the data set and the representativeness of the firms in the sector seem to offer solid and reliable findings.

5.6 DEVELOPMENT OF SCORECARD AND PERFORMANCE CRITERIA

In the next paragraphs the development of the innovation success criteria will be explained. The procedure to develop these measures included the following steps:

1. performing a literature search on success criteria;

2. analysing empirical findings of the original study with respect to success measures;
3. determining the relevance of that criteria for this ICT study;
4. selecting commonly used criteria from empirical studies and comparing these with the original set of criteria;
5. including a complete set of variables in the benchmark survey;
6. limiting 25 variables to a valid, workable set of 22 success variables;
7. correlating success variables with company data.

Literature search on success criteria
The literature on new product development continues to grow, leading to a variety of studies on measuring success of product development. Brown and Eisenhardt (1995) indicated three innovation research streams (see also chapter 2). They pointed out that each stream centred on a particular form of performance. The rational plan research especially concentrates on financial performance measurement (profits, turnover, market share). The communication web research especially addresses the effects of communication on project performance, using perceptual success measures (team and management ratings). The disciplined problem-solving stream focuses on the product effects on the development process, by using operational success measures (e.g. speed and productivity).

In their report on measuring product development success and failure, Griffin and Page (1996) observed that success is a *multidimensional* concept, i.e. that new product success can be considered in different ways. Companies can innovate for different goals. Some companies innovate to achieve a better efficiency, or a larger market share, while others want to enter new business areas or improve turnover and profitability. Clearly, innovation has a multidimensional flavour, just as innovation success.

Innovation success measurement: some methods
Cooper and Kleinschmidt (1995) conducted a multifirm benchmarking study to identify the company-level drivers of success in new product development. Given the multidimensional nature of NPD performance they used 10 performance measures of a company's new product programme: a success rate, percentage of turnover, profitability relative to spending, technical success rating, turnover impact, profit impact, success in meeting turnover objectives, success in meeting profit objectives, profitability relative to competitors, and overall success. Based on factor analysis of the metrics they constructed two overall dimensions: *programme profitability*

and *programme impact*. Programme profitability comprised the company's innovation programme profitability rating and the overall success rating, both relative to competitors; whether the programme met profit objectives; the programme's profitability relative to spending; the impact of the programme on firm's profits; and whether the programme met turnover objectives. Programme impact comprised the percentage of turnover by new products; the impact of the innovation programme on company turnover and profits; an overall success rate, and a technical success rating. A similar procedure was performed in Cooper's study (1996) of 161 firms on the drivers of new product performance in business units within corporations. By using this procedure, insight was gained into the factors and practices that discriminate between top and poorer performers. Ten key performance metrics were used identical to the 1995 study. Based on the factor analysis two success dimensions were constructed: *profitability* and *impact*. Profitability captures how profitable the business's total new product efforts were. This dimensions consisted of profitability versus competitors and overall success rating of the business's total new product effort; whether the total initiative met profit objectives; its profitability relative to spending; and the impact of the total effort on the business's profits. Impact was concerned with the impact of total new product efforts on the business. This performance dimension consisted of percentage turnover by new products achieved by a firm (in their study a business unit); the impact of new products on both turnover and profits of the unit; the success rate achieved; and the technical success rating.

Cooper & Kleinschmidt (1987) conducted research on different dimensions of new product success based on a sample of 125 companies, with 123 new product successes and 80 failures. They found three independent dimensions of success: *financial performance, opportunity windows,* and *market share*. These dimensions are a reduced subset of ten measures of new product performance: profitability level, the product's payback period, domestic market share, foreign market share, relative turnover, relative profits, turnover versus objectives, profits versus objectives, opportunity window on new categories and opportunity window on new markets. The financial performance dimension captured the overall financial success of the product. This comprised of relative profits, meeting profit objectives, meeting turnover objectives, relative turnover, profitability level, and payback period (negative). The opportunity window item shows the degree to which the product/service opened new opportunities for the firm (in terms of category of products and a new market for the its products). The third dimension, market share, describes the impact of the product on both domestic and foreign markets. This dimension has been constructed by using market share measures and relative turnover and meeting turnover and profit objectives.

In a large sample benchmark study including 135 firms (Cooper & Kleinschmidt, 1995) new product performance was captured using seven measures of company's new product programme performance. New product programme referred to the totality of new product efforts of the company or division of analysis. Firms from the USA, Canada, Germany and Denmark participated. This study explored the relationship between practices and performance by considering the impact that each practice or firm characteristic had on performance. In the research design seven metrics were defined: meeting company sales objectives, profit objectives, success rate of new products, sales impact of entire programme on company, profit impact on the company, profitability relative to competitors and overall success, as a global measure relative to competitors.

Overall, these studies have provided useful insights into how new product success can be measured and which independent and unique dimensions of success can be used. Given the fact that many different concept of success were used in the past, the financial return on investment is one of the most easily quantifiable parameters (Maidique & Zirger, 1985). At the same time it is not the only important one. Success is clearly more than the financial benefits accruing from innovation. More extensive studies, such as NewProd II (Cooper & Kleinschmidt, 1987), incorporated more measures, with more explanatory power. Absolute figures, like profitability levels, were no longer measured in absolute terms but related to competitor profit figures. Profitability as a level of success was constructed relative to meeting project objectives, and management-rating measures were used in addition to more quantifiable financial measures. In this way success was measured in different directions.

5.7 INNOVATION SUCCESS CRITERIA

Building upon the above methods we developed a benchmark instrument. Among the variables in this survey were 25 success variables. The variables were constructed and derived on the basis of an extensive literature search on organisational effectiveness and effective innovation management (Griffin & Page, 1996). The criteria for the selection of the variables related to the validity, rigour, and relevance of the success variables in earlier empirical studies. When constructing the set of success variables, emphasis was placed on measuring existing structure, process and action outcomes. In other words, the study tried to assess performance from realistic, observable sources.

First, data results revealed that most firms performed well. The positive growth figures, on many success measures, reflect the growth development of the sector. When we relate turnover and profit growth to innovation the following results appear (table 5.2). On average, new products and services introduced throughout the past three years accounted for 38% of annual turnover and 39% of annual profit.

Table 5.2: Contribution to turnover and profit resulting from innovation

Contribution to:	Turnover in %	Profit in %
1. Products & services newly introduced in the market	38%	39%
2. Products & services drastically improved during the past 3 years	30%	30%
3. Products & services hardly improved during the last 3 years	32%	31%

With respect to performance in the development process, an average of 39% of new product and service ideas were selected to become an innovation project. 56% Of the projects succeeded in reaching the commercialisation phase. This implies that on the basis of this study sample of 32 firms, on average one out of five ideas, resulted in a commercial product or service in the ICT sector.

Innovation scorecard: the measurement
The measurement of innovation performance has a multidimensional nature. We considered innovation success in different ways and included seven measures of company's innovation performance:

1. process success;
2. portfolio success;
3. growth success;
4. opportunity success;

5. project success;
6. quality;
7. overall innovation success.

Each of these seven performance dimensions and the rational for their inclusion in the framework are highlighted below.

For an extensive statistical interpretation of the measurement factors the reader is referred to Huizenga (2001). The scorecard is developed to identify the major drivers for success. This framework was based largely on previous research findings as well as on the literature discussed in this book. The framework has been replicated in several studies (Huizenga, 2001; Cobbenhagen 1999; Cobbenhagen, Den Hertog, Pennings, 1995).

Factor analysis

The next step in the performance evaluation was to construct a performance map to reduce the individual performance variables (Cooper & Kleinschmidt, 1995). This was done through a factor analysis. The purpose of factor analysis was to deduct a set of underlying, unobservable patterns (factors) from covariance relations among individual variables. The variables within one factor show a high correlation, whereas they are hardly correlated with variables in another factor. Several criteria have been used for the final choice of the number of factors (Cobbenhagen, 1999):

*   the proportion of variance explained;
*   knowledge of the subject matter;
*   the 'reasonableness' of the results;
*   the goal to obtain a simple structure of factor solutions.

Factor analysis technique was used to derive a comprehensive set of success variables. Innovation success is multidimensional phenomenon. Therefore we used a factor analysis to obtain a range of success criteria. The statistical procedure included the following steps.

1. Based on the extensive literature survey and the validated MERIT-benchmark questionnaire (see: Cobbenhagen, 1999) 25 individual success variables were selected. Before applying the survey we limited the set of 25 success variables to a range of 22 success variables. This was based on the applicability findings of an ICT expert panel. Due to data availability a workable set of 20 success variables was used (table 5.3 provides the details). The research also aimed to find a simple structure of the factor solutions.

2. A principal component analysis was conducted, with a sample of 32 companies (n = 32): based on the results we extracted 6 factors on the basis of an Eigenvalue larger than 1. This set of factors explained 78% of the variance. The correlation matrix indicated that the variables were suitable for factor analysis. The KMO statistic was 0.61, which is 'middling appropriate' for factor analysis (Kaiser, 1974).

3. A Varimax rotation was conducted: we used this rotation technique to converge to better success variables, which are better interpretable. The results of this Varimax rotation are presented in table 5.3. The advantage of performing Varimax rotation is a higher correlation of the individual variables with the factors. We observed that 5 out of 20 variables were correlated to more than one factor, indicating that some factors are somewhat correlated to each other. This is especially the case with the following variables: rapid product & service introduction, continuous innovation, capturing new market opportunities and cost reductions. Although a small violation against simple structure analysis (not all variables perfectly correlated with one and only one factor), the results produced an interpretable set of factors.

4. Based on the statistical results and grouping of variables we labelled the factors as process success, portfolio success, growth success, opportunity success, project success and quality.

5. To rate the overall success of a firm the top management and technical managers were asked to rate their firm's overall performance. Intercorrelation of the overall success measure and the factors reveals no significant intercorrelation. This overall success measure was used together with the six other extracted success measures.

Table 5.3: Factor loading on success variables

Variable	Process Success	Portfolio success	Growth success	Opportunity success	Project success	Quality
Turnover growth from new products	0.45					
Introduction new process technologies	0.67					
Degree of technical creativity	0.75					
Technological competence	0.85					
Technological knowledge transfer	0.85					
Cost reductions	0.60		0.45			
Capturing new market opportunities	0.63		0.57			
Rapid product & service introduction	0.40	0.49	0.42			
Continuous innovation	0.43	0.59				
Turnover from new products and services		0.91				
Profit from new products and services		0.92				
Turnover growth			0.76			
Profit growth			0.62			
Growth of employees			0.86			
Turnover growth by entering new markets				0.87		
Profit growth by introducing new products and services	0.42			0.50		
Profit growth by entering new markets				0.90		
% of new ideas turning into projects					0.80	
% of projects that reach market introduction					0.81	
Degree of product & service quality						0.85

Notes
Factor loadings less than 0.4 have been omitted for readibility.

Table 5.3 (continued): Short description of success variables

Variable	Short description	Rating
Turnover growth from new products	The degree to which a company, compared to its competitors, performs in the realisation of turnover growth by introducing new products into the market	Managerial rating; Likert-type scale, score 1-5*)
Introduction new process technologies	The degree to which a company, compared to its competitors, performs in the introduction of new process technologies	Managerial rating; Likert-type scale, score 1-5
Degree of technical creativity	The degree to which a company, compared to its competitors, performs on technical creativity	Managerial rating; Likert-type scale, score 1-5
Technological competence	The degree to which a company, compared to its competitors, performs in technological competence	Managerial rating; Likert-type scale, score 1-5
Technological knowledge transfer	The degree to which a company, compared to its competitors, performs on technological knowledge transfer through co-operation agreements	Managerial rating; Likert-type scale, score 1-5
Cost reductions	The degree to which a company, compared to its competitors, performs in cost reductions because of new technologies or materials and tools	Managerial rating; Likert-type scale, score 1-5
Capturing new market opportunities	The degree to which a company, compared to its competitors, succeeds in capturing market opportunities	Managerial rating; Likert-type scale, score 1-5
Rapid product & service introduction	The degree to which a company, compared to its competitors, performs in rapidly introducing product and services into the market	Managerial rating; Likert-type scale, score 1-5
Continuous innovation	The degree to which a company, compared to its competitors, performs in continuous product and services innovation	Managerial rating; Likert-type scale, score 1-5
Turnover from new products and services	The contribution to turnover from new products and services	Financial rating, % of total turnover
Profit from new products and services	The contribution to profits from new products and services	Financial rating, % of total profit
Turnover growth	The realised turnover growth 1996/1995	Financial rating, growth rate
Profit growth	The realised profits in 1995	Financial rating, growth rate
Growth of employees	The realised growth in number of employees 1996/1995	Financial rating, growth rate
Turnover growth by entering new markets	The degree to which a company, compared to its competitors, performs in the realisation of turnover growth by creating new markets	Managerial rating; Likert-type scale, score 1-5
Profit growth by introducing new products and services	The degree to which a company, compared to its competitors, performs in the realisation of profit growth by introducing new products into the market	Managerial rating; Likert-type scale, score 1 to 5
Profit growth by entering new markets	The degree to which a company, compared to its competitors, performs in the realisation of profit growth by creating new markets	Managerial rating; Likert-type scale, score 1-5
% of new ideas turning into projects	The actual percentage of new product ideas that resulted in a project	Percentage rating
% of projects that reach market introduction	The actual percentage of innovation projects that reached the market and were successful	Percentage rating
Degree of product & service quality	The level of quality of products and services	Managerial rating; Likert-type scale, score 1-5

Notes
For more detailed information: the full questionnaire is available upon request (Huizenga, 2001).
* Scale:
1 = company lags behind the pack
2 = company drives in the end of the pack
3 = company drives in the middle of the pack
4 = company leads the pack
5 = company leads the frontrunners group

The statistical interpretation and grouping of the variables is argued below.

1. *Process success*: This performance dimension consists of seven success variables. All of them were management ratings. The ratings were closely correlated with each other. This dimension puts emphasis on: (a) an innovation's contribution to turnover growth, and (b) the incorporation and translation of new technologies into new processes. These measures have a strong correlation. This might be an indication that companies, who have been successful at absorbing advanced technologies and turning these into new products and services, are also high growth companies, in terms of turnover. This performance dimension includes the following success variables:

   - realisation of turnover growth by introducing new products on the market;
   - successful introduction of new process technologies;
   - the degree of unique technical creativity;
   - the degree of technical competence relative to competitors;
   - achievement of cost reductions resulting from new technologies, resources, and tools;
   - successful transfer of technological knowledge through co-operation;
   - capturing new market opportunities.

Brown & Eisenhardt's research (1995) of the empirical innovation literature supports the underlying rationale for these dimensions. The combination of the two factors, process performance and turnover-related success, might be the following. Process performance entails the introduction of new technologies (e.g. Java, HTML), resources (e.g. new technical know-how) and tools (e.g. software development tools). A productive process means lower costs and hence lower prices for products and services, which in turn might lead to increased product success. Furthermore, the introduction of new technology creates a faster process and might lead to more flexibility and less time for development and production. This may result in greater financial success, in terms of turnover growth. The process success dimension belongs to the communication web stream of innovation research as discussed by Brown & Eisenhardt (see chapter 2). It consists of perceptual success measures provided by multiple informants, e.g. management ratings of overall technical performance and success in terms of achieving turnover growth and cost reduction.

2. *Portfolio success*: This performance dimension consists of four success variables. Two of them are a management rating, the other variables are percentage ratios. Such ratios should be a better measure than an absolute number due to the varying sizes of the business units and firms in our sample. This factor portrays the success of continuous innovation. Successful firms that, relative to their competitors, pick up innovative ideas before their competitors do so and transform them quickly into new products and services. These firms are the first to seize profit and 'turning good ideas into money quickly'. The factor includes:
    - continuous innovation in products and services;
    - the speed of innovation, e.g. rapid product and service introduction;
    - the percentage of turnover generated by products and services introduced throughout the past three years (see also table 5.1);
    - the percentage of profit that generated by products and services introduced throughout the past three years (see also table 5.1).

This performance dimension relates to the rational plan research stream in terms of Brown & Eisenhardt (1995). Such measures have been used in other studies to measure performance where the latter two variables are considered core success measures and a standard success measure used by the Product Development Management Association (PDMA) (Griffin & Page, 1996; Barczak, 1995). It characterises new product effectiveness, whereas the former dimension emphasises process performance. Portfolio success is primarily oriented towards the financial performance of new products and services relative to the existing range of products and services. This dimension is similar to Cooper's (1986, 1995) success dimension of impact. The portfolio success measure takes the impact of innovation efforts on the business into account.

3. *Growth success*: This performance dimension consists of turnover, growth & profit success variables, both measured as percentage ratios. It includes:

- the turnover growth, of each company for the years 1996 and 1995;
- employee growth, of each company for the years 1996 and 1995;
- profits (as a percentage of annual turnover) of each company for the years 1996 and 1995.

This performance dimension is a financial measure that closely relates to the ration plan research stream. It characterises how profitable the business of the total new product and service efforts and process innovations are. This dimension mirrors success ratings at the company level, used in the SAPPHO study (1972, 1974) and many of Cooper *et al.*'s NewProd studies (1979, 1986). In a replication study of NewProd (Cooper & Kleinschmidt, 1992), using data from 123 industrial product firms, success and failure were defined from a financial viewpoint. The study used the degree to which the new product's profits exceeded (or fell short of) the firm's acceptable profitability level for this type of investment. The constructed performance dimension is identified (Griffin & Page, 1996) as an appropriate measure.

4. *Opportunity success*: This performance dimension consists of three success variables, measured by management ratings. It is a factor that describes the success of new business development efforts. It consists of:
   - The degree to which a company, compared to its competitors, performs in the realisation of turnover growth by creating new markets;
   - The degree to which a company, compared to its competitors, performs in the realisation of profit growth by introducing new products into the market;
   - The degree to which a company, compared to its competitors, performs in the realisation of profit growth by creating new markets.

This dimension is similar to what Cooper & Kleinschmidt (1987) identified as an 'opportunity window'. Business opportunity is the degree to which the product or service opened new opportunities to the firm resulting in a new product category and a new market for the firm. The dimensions adhere closely to Hamel & Prahalad's (1994) idea of stretching a firm's competence endowment to pursue attractive opportunities. The dimension of business opportunity belongs to the communication web research stream (Brown & Eisenhardt, 1995).

5. *Project success*: This performance dimension consists of two success variables, both measured as percentages. The 'project success' measure describes the success of the project efforts of the firm. The degree of success refers to (a) the generating of new product/service ideas, and (b) the project selection mechanism of the firm to select ideas and turn them into projects. Furthermore it portrays the success of managing innovation projects in commercially launching a new product or service idea. This factor consists of:
   - the percentage of new ideas that have become a project;
   - the percentage of projects that have resulted in a product or service that is launched in the market.

Cooper & Kleinschmidt (1995) used a similar measure, which they called success rate, which measures the percentage of projects that entered the development stage and were ultimately considered a commercial success. Both Iansiti (1994, 1995) studies of 27 development projects in the mainframe industry and Eisenhardt and Tabrizi's (1995) study of 72 development projects in 36 computer firms used similar project-related success measures. These studies used speed (of the development process) as a performance measure. This measure portrays the success of the development efforts. Both studies related this performance dimension to such factors as pre-development work and early project planning. The project success criteria fit with the disciplined problem-solving research stream (Brown & Eisenhardt, 1995). This stream is concerned with problem solving in projects (e.g. by cross-functional teams) and high communication flows, to successfully execute the project step in the development process.

6. *Quality*: This performance dimension consists of one success variable, which was rated by the director. Single-item measures are not generally considered 'good measures' of any construct. However, the measure was strongly independent of the other success measures. This variable describes the quality of the business offerings of a company, as observed by the director. It consists of the variable: quality of products and services, relative to competitors.
This dimension characterises the successful adaptation of products and services to customer needs, thereby realising high-quality product/service offerings. This is often regarded as a pre-requisite for

the firm's continuity. Striving for quality improvement is an important factor in realising competitive advantage. Keller (1986) used a quality measure in a study of 32 R&D project groups. In this study, quality was related to the quality of the project outcome. Clark & Fujimoto (1991) and Womack *et al.* (1990) used, among speed and productivity variables, quality as a performance variable in their in-depth case studies in the automobile sector. The quality performance dimensions mostly fits with the disciplined problem-solving research stream, which is characterised by single-industry studies (Brown & Eisenhardt, 1995). In this stream, quality is often viewed as a high-quality product/service concept resulting from a fast development process.

In sum, using performance dimensions that can be described along the three streams of innovation research proposed by Brown & Eisenhardt (1995, p. 365), we focused on somewhat different aspects of innovation success. Complementary success variables have been combined with independent factors in order to cover the many dimensions of innovation success.

Overall innovation success
The constructed performance dimensions are independent. A final step in the performance evaluation was the use of an overall success measure. Next to these success dimensions, top management and technical managers were asked to rate the overall success of their firm's innovative efforts. Intercorrelation of the overall success measure and the factors reveals no significant intercorrelation. This provides strong support for the relevance and validity of the overall success criteria. With these categories of success measures we continued further statistical analysis to identify the key success factors in innovation. What then are the key success factors in innovation, the factors that drive performance in IT companies?

Finally, a reservation should be made in advance as to the significance of the correlation coefficients in this study. On average, this study reveals no highly significant findings. Amongst the possible explanations are the sample size, the method of performance indicators, and the sector context e.g. strong economic growth. However, the data set is large in terms of number of variables, data points and interview respondents. This might suggest that the data set is normally distributed, which might explain the lack of exceptionally high scores for the variables. Nevertheless, the study results and correlation coefficients provide an idea of the most interesting relationships between innovation performance and firm characteristics.

PART III:

The Management Practice

6. Strategy and Innovation Performance

6.1 INTRODUCTION

This research provides insight in to how companies fare in terms of certain innovation practices and whether or not the practices really matter. Do they impact performance? Consider now some of the complications for better innovation management, which evolved from the study findings.

In the early 1990s, the ICT sector was facing an image problem due to the lack of growth, unreliable hardware suppliers, the effect of an economic slowdown, clients freezing their ICT expenditures and a questionable quality and service level of many suppliers. All of these developments had a damaging impact on the sector and resulted in companies slimming. Companies disappeared altogether and there were a lot of take-overs between hardware, software and service providers. Later in the 1990s, the economic tide changed. Suppliers became more known as reliable business partners, ICT expenditures were on the rise, and the higher quality of personnel contributed to an improved sector image. New business lines emerged, such as ERP implementation (enterprise resource planning), CRM (customer relationship management) consulting and e-commerce.

One of the factors causing these growth rates in the late 1990s was the higher demand for client/server solutions in the hardware sector. Increased quality, speed, reliability and performance of computers also boosted hardware sales. Firms were investing in systems and networks and became more confident about relying on ICT for their business processes. The services sector profited from the increased demand for consulting, facilities management, system and network implementation, and operation services. Firms invested more in these areas of ICT in their search for more effective business processes to get ahead of competitors.

The past Millennium and Euro conversion projects led to steeply rising turnover figures of ICT companies. Before the twentieth century came to an end, computers had to be adjusted to deal with the change in date. Inventory systems, distribution centres, reservation systems, process monitoring systems etc., had to be adjusted in order to prevent systems downtime.

The most recent development in the ICT industry was the emergence of the Internet and e-commerce. Business-to-consumer and business-to-business services and markets provided new roles due to ICT. ICT was connecting persons, consumers and business enterprises throughout the world, irrespective of time, geographic or organisational boundaries to form a 'network area'. ICT extended the reach of every individual to communicate and share information leading to new business ventures.

In this hectic business context of the beginning of 2000, the research study identifies a number of interesting insights in the ICT sector that are not explicitly addressed in the innovation literature.

6.2 STRATEGY

Innovation can be a matter of strategic importance for companies. It is a competitive weapon to expand the core business of an enterprise. This strategic importance is reflected in attaining market share, outclassing competitors and attracting new customers. In this context, the companies' strategic vision of innovation tells us something about the firms' aspirations. In particular, we looked at innovation awareness and strategic commitment of companies.

In our interviews with over 100 CEOs, business development managers, marketing and ICT managers we recorded a great number of statements about strategy. The research questions on business strategy, core competencies, portfolios and competitive advantage hardly needed additional explanation in the interviews. Strategy is a hot item in the ICT sector. Even more, the interviewed managers anticipated many strategy questions. The interviewed persons seemed well trained in using the concepts of strategy and competitive advantage.

The CEO of an innovation unit of an ICT solutions provider remarks on their business strategy:

> Our unit was set up together with the introduction of market divisions. It's the
> result of a great deal of corporate strategic discussions that have been put on

paper in different flavours. The strategic goal of our unit is to refresh the portfolio of the divisions.

Next to portfolio ideas, core competence ideas were among the management talk. The business development manager of a European ERP software company remarked:

> The management is aware that the company's strategic focus is with new product development... That is our core competence... .

The vice-director of an ICT system builder expressed a similar strategic vision:

> Our firm is in the middle of making a strategic reorientation. We want to transform from an application builder to system components and assembler of applications. New product - market combinations by stretching our strategy and doing the right things right ... .

Value added thinking was often observed in the interviews. The CEO of a large software developer explained:

> Our business is a high-value software product. High-tech is not handcraft labour. Our strategic vision is based on an awareness of industrial thinking.

The managing director of a large system builder, part of a European corporation:

> The stage of the product life cycle of our market is decline. Our company aims to step out of the business before actual decline takes place and step into new business. The strategy formulation for the next 5 years is determined: business formula and mission first, then positioning of core operations within the corporate enterprise.

All these statements lead one to believe that the firms and their management teams have thought through their strategy, positioning and competition on capabilities. One might expect differences in strategic focus between firms, particularly if managers paid attention to their own business strategy and to differentiating from competitors. On the other hand, the dominance of strategic talks by managers might raise the suspicion that there is a difference between what is said and what is actually true. Remarkably fewer stories included the actual strategy implementation and the balance between 'strategy in theory' versus 'strategy in action'. With this research lens in mind we are looking at the data presented here.

The first interview question (What does innovation mean to your company?) aimed to identify the awareness in (top) managers of the relevance of innovation. Given the intensive strategy discussions that were recorded a clear vision of innovation could be expected. Let's give some anecdotal case examples that firms have different visions of what innovation actually is.

The CEO of a firm specialised in system control, and part of a large international ICT service provider reveals:

> Product innovation to us means the ability to constantly translate feelings in the market into services that are targeted to that market.

The director of an ERP software producer and market leader in a profitable market niche gives another opinion:

> For us, the real renewal lies in two areas of process innovation: to transform to object-oriented development and the maintenance and renewal of existing software according to rapid application development (RAD) method.

A business unit of a large ICT service company, a software producer and system developer remarks:

> Innovation is making new software development environments and inventing new products.

The manager of a system developer believes they are unique:

> Innovation has been given shape in a different way in our company than in other companies. Others do not pay attention to process innovation. Our vision and approach of system development is based on object orientation.

But the CEO of another company, a large facilitation service provide tells a similar story:

> Being one or two steps ahead of the customer's developments. Understanding the business processes of the customer even better than they do. That is how we try to innovate ... .

Finally, the interim managing director of a software producer and market leader for the semi-governmental market puts the customer in front:

> The core of innovation is to offer services to customers that bring them advantage in terms of time, money and insight. The technology is a side aspect.

Let's see how these firms actually differ in their strategic priorities, innovation style and resource commitments to innovation.

Management priorities

Managers have to make choices. These choices are reflected by the policy priorities of the firm made over a longer period of years. Measuring the policy agenda at two successive time periods might identify the path dependency of the firms. Priority setting for successive time periods was used as a way of acquiring a clear indication of the companies' path dependency, e.g. the awareness and choices regarding innovation. In the survey these strategic priorities were measured, in 1997 and 1998, for two time lags: 1988 to 1992 and 1992 to 1996. In the competitive ICT markets it is more difficult for managers to set priorities. More complex technology issues have to be addressed at the same time. We assumed that ranking a set of business issues might suggest how much attention is actually devoted by management

Innovation strategy style

Secondly, we tried to find out if innovation was also expressed in the company's business strategy. The survey presented a set of questions aimed at labelling the strategy of the firm. In this way, the study looked for a more detailed characterisation of the innovation dimension in the business strategy. Based on the survey we provided CEOs and managers with a set of strategy descriptions. Labels such as creative inventor, first mover and technology imitator have been used to indicate the firm's strategic focus on new markets, technologies or competencies.

Resource capacity commitments

Thirdly, we raised the question as to whether this commitment to innovation was actually implemented. The realisation of innovation efforts requires the allocation of such resources as time and money. The survey included resource criteria to indicate the actual commitments to innovation. In other words, it indicated how much money and time was (structurally) spent and to which activities.

6.3   STRATEGIC PRIORITIES ON THE MANAGEMENT AGENDA

Results

To provide insight into the management agenda, the respondents were asked to mark their policy priorities. Twenty-five topics were presented with a 1-5 Likert-type scale ranging from low attention to top priority. Figure 6.1 provides the descriptive data for the most important topics. As we observed previously, strategy is a core discussion topic in the ICT sector. What then is its impact on the management agenda? In retrospect the data reveal that the management agenda has drastically changed over time. The subjects that were high on the agenda years ago have been replaced by new topics such as a focus on core business and quality. The topics that attracted little attention in the period 1988 to 1992, such as training, education and quality improvement have been given reasonable to high priority more recently (1992 to 1996). The most urgent topics attracting management attention are quality assurance, profit improvement, recruitment of personnel and concentration on core activities. We will discuss these into more detail now.

Concentration on core activities

This topic has been given the highest priority by the majority of ICT companies, which suggests that software producers and service providers have fundamentally (re)discussed their own business. Interviews reveal that companies have investigated their value chain to identify those activities that generate most added values. Also activities outside the core portfolio have been assessed in terms of their synergistic advantages. This movement of making organisations 'lean and mean' has resulted in throwing away 'excess weight'. Business activities that do not directly contribute to strengthening the core business are abandoned.

Enterprise growth

Expanding current business has also been a high priority for both software developers and service providers. The management of ICT companies has put growth-related matters high on the policy agenda. Throughout the last ten years, growth of turnover has been given reasonable to high priority. It was the most important policy matter during the 1988 to 1992 period. Turnover growth is regarded as a key management objective that attracts constant attention. The increase in market share continues to be a dominant item for expanding a company's business. Directly related to enterprise growth is personnel. The shortages of ICT labour during the late 1990s supply are a dominant restrictive factor for further growth of the companies, resulting in a relatively high priority for personnel recruitment. Another key priority at the top management level is an increase in employee productivity. Due to the rapid growth, there was less attention for efficiency, leading to increasing production costs.

Figure 6.1: Strategic priorities for ICT companies

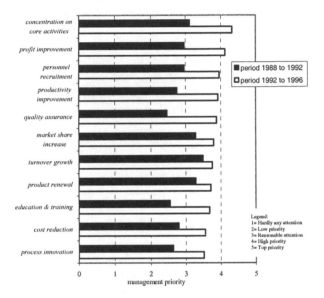

Competitive pressure and the need for fast and high-quality product and service delivery might explain the increased need for more efficient working methods and productivity increase.

Innovation
One remarkable finding is that innovation is not among the top three priorities of the management of ICT companies. Product renewal is the first innovation-related aspect and one of the important topics mentioned. In the 1992 to 1996 period, process innovation (especially in ICT implementation and advice processes) attracted more attention. The item shifted from the range of reasonably important to high-priority matters. A reverse 'path-dependent' movement can be found for product renewal. Product renewal was among the highest priorities in the early 1990s. Product renewal as such

was not one of the hot topics on the agenda. The attention for growth and profit improvement was of most concern for top managers.

Quality assurance

This topic gained several places on the management agenda as compared to a number of years ago. The attention attached to quality by the top management of ICT firms indicates that it is a dominant subject in today's ICT markets. Quality assurance had a low priority in the period 1988 to 1992, but is among the high priority matters for management today.

Business-specific priorities: producers versus service providers

We were also interested to see whether the data suggest a distinction between the policy agendas of software producers and service providers. It appeared that software producers specifically focused management attention, for the period 1988 to 1992, on:

- product renewal, e.g., new server platforms based on new technologies;
- turnover growth, e.g., partnering with service providers;
- increase in market share, e.g., take-over of smaller competitors.

Remarkably, top priority was given to the following subjects in the period 1992 to 1996:

- product and process innovation, e.g. a decision by a large company to set up an independent innovation unit;
- productivity improvement, e.g. a training programme for Java software developers;
- concentration on core activities, e.g. a developer of transaction software decides to withdraw from billing services.

The strategic subjects for ICT service providers are similar to those of software producers for the period 1988 to 1992. In particular, the focus on turnover growth and market share increased. However, the policy agenda of service providers in the 1992 to 1996 period dramatically changed and focused on three subjects, listed in order of priority:

- recruitment of personnel, e.g. a hardware maintenance company started recruiting higher vocational education (students);
- core activities, e.g. a software maintenance company withdrew from PL1 programming;

- profit improvement, e.g. a software developer made low return on investments.

These variables might explain a clear distinction in strategic focus between producers and service providers. Software producers tend to focus more on innovation, whereas service providers concentrate on growth and result improvement. As the strategic discussions in the ICT sector are thriving, these findings do not differ from our expectations that companies differ in strategic focus. The next question to be raised, of course, is which management choices and actions really matter for excellent innovation performance.

6.4 KEY SUCCESS FACTORS IN STRATEGIC PRIORITIES

Table 6.1 shows the correlation coefficients of the strategic priorities and the innovation success criteria. The table reveals the correlation between variables, but reservations need to be made because no high correlation levels were found.

The results indicate that ICT firms, which gave top priority for quality assurance, are among the frontrunners in innovation success. Firms, which gave strong priority to training, process innovation, technical co-operation and protection of intellectual assets, differentiate themselves by excelling in managing the innovation process. A few firms gave priority to profit improvement, turnover growth and expansion of distribution channels. These firms were successful in finding new opportunities and commercialise them.

Table 6.1: Correlation coefficients between strategic priorities (period 1992 to 1996) and innovation performance

Innovation scorecard: Individual success items:	Overall success	Process success	Portfolio success	Growth success	Opportunity success	Project success	Quality
Concentration on core activities	0.21					0.30*	
Profit improvement				-0.37*	0.28*		
Productivity growth				-.35**		0.29*	
Turnover growth					0.28*		
Quality assurance	0.33*						0.46**
Product renewal			0.44**				
Training		0.28*					
Process innovation		0.39**					
Expand distribution channels					0.24*		
Technical co-operation		0.38**		-	0.36**		
Protection intellectual assets		0.32*					

Notes

Strategic variables for the period 1992 to 1996 are listed in order of decreasing average priority (* denotes significance at the 0.05 level and ** denotes significance at the 0.01 level). Significant correlation coefficients in the range 0.3 - 0.4 are shown and remarkable correlation results.

6.5 DISCUSSION: STRATEGIC PRIORITIES

At the beginning of this chapter we implied that there is a difference between what is said and what is actually true about strategy in ICT firms. In this section, we will see that many firms acknowledge the strategic relevance of quality and core competencies and at the same time that firms differ in their timing of strategic priorities.

Quality assurance
Quality assurance has become a dominant item in the business operations of ICT companies. The argument for this high priority setting may lie in the belief that companies view quality as a differentiating factor. Equally relevant is the fact that the ICT sector has for many years suffered from an image of projects that went over budget and time deadlines. In order to

improve their image and operations, a lot of ICT firms have started to articulate the quality of the product and service offerings. The attention for quality matters is found in the increased dedication to service level agreements as a means of ensuring the delivery of services as promised.

The slogan of Company E is 'continuity and quality, your partner'. Companies using IBM hardware were often tied to expensive IBM service contracts. However, competitor of IBM, Company E identified a new market space for a new form of services: third-party maintenance. The standard service contracts of IBM were laid on the table, which led to the rise of a new market for maintenance. Companies including Olivetti and Digital Equipment and company E became each other's competitors. Throughout the years, the standardisation of software and hardware had their influence on the development of the maintenance market. Quality of service and personnel were the important selling arguments for company E. Quality became their leading motive for the delivery of multi-vendor services. As the reliability of systems and software increased, the demand for hardware maintenance appeared to be less. Company E realised that an adaptation of their service portfolio was needed, and that quality required a new form and dimension. Its Service Reengineering Project was intended as a redefinition of the business formula. In the first phase, the quality of activities was assessed by means of:

- a customer satisfaction survey;
- in-depth client interviews;
- a lost-order analysis.

Ensuing from this project were new quality services referred to as 'integral services management'. A key feature was to aim at the highest level of quality in maintenance and control processes of automation. This existed of ITIL (Information Technology Infrastructure Library) control elements including:

- service level management;
- availability management;
- capacity management.

In various instances, firms indicated that quality assurance is a spearhead and integral part of their operations. Just like company B:

Company B focuses on 'Excelling with people'. In order to secure the quality of personnel posted at their clients, company B wanted to prepare the way for internal knowledge sharing. After all, the personnel had to gain more knowledge of client server platforms. A knowledge bank 'Transform' was designed, which included:

- techniques and methods to control and steer Client/Server projects;
- information on Rapid Application Development with example documents, techniques, checklists and risk management techniques.

Company B had developed an indirect steering instrument to increase the quality and productivity of its employees.

The focus on quality was a returning and explicitly stated element for ICT firms in striving for highly qualified people and high-quality products and services.

Concentration on core activities

The results show that 'concentration on core activities' is moderately correlated with overall success (r=0.21). This variable is significantly correlated with project success (r=0.30*). The ICT study confirms that companies pay attention to streamlining and focus on the core activities.
On the basis this outcome, one might argue that ICT companies that have concentrated their business activities are also more successful:

- at selecting potentially profitable new product ideas;
- at turning innovation projects into commercial viable products and services.

This indicates that a firm restricting itself to its core business activities also carefully selects new product ideas that fall within the range of the core business. This means that firms developed products and services that did not take the firm into new and unfamiliar markets, nor required technology that was totally new for the firm.

A next set of key success factors relates to priority for turnover growth, profit improvement and expansion of distribution channels. Firms that prioritise these matters are particularly successful in generating profit and turnover from entering new markets (opportunity success). Process performance is most related to technical co-operation and management attention for process innovation. Companies that have directed attention to improvement of operational processes and invested in strengthening technical partnerships show operational excellence, just like software development company F:

The manager is head of business company operations. His responsibilities include:

- Technology & Business Development, including research & development and finding the right product market combinations.
- Business operations, ensuring that the right people are on the right job, the right proposal are developed for clients and the internal automation is improved.

In this context, project PIAG was started to improve the operational project control and planning of software projects. The goal was to set up a generic working method for system development. PIAG had to ensure one common operation, irrespective of the technical development environment. This implied the generation of system solutions was not based on one specific software programming language. At the same time, company F was in the midst of a process involving the switch from mainframe solutions to the development of client server solutions. Project PIAG resulted in tying up with partners to ensure the Cobol and Oracle knowledge needed for their service delivery.

Bottom line conclusion is that managers who *deliberately* prioritise process innovation excelled in managing the innovation process.

Strategic priorities in retrospect: 1988 to 1992
To provide an indication of the changes in strategic priorities over a longer period of time, the same variables have been measured on the time axis, covering the years 1988 until 1992. We explicitly asked managers to look back in time to prioritise policy items. The data can provide insight into how managers perceived changes in strategic priorities. To some extent it also provides, in retrospect, a kind of pattern in the strategic priorities and the priority setting of ICT firms, based on managers' perceptions (cf. Van de Ven & Poole, 1990).

On the basis of the correlation analysis we are able to make a number of remarkable observations. First, the success factors differ. In the late 1990s quality assurances was a key success factor, whereas low attention to acquisitions and productivity growth were keys to success in the early 1990s. Table 6.2 gives the correlation results.

Table 6.2: Correlation coefficients between strategic priorities (period 1988 to 1992) and performance

Innovation scorecard: Individual success items:	Overall success	Process success	Portfolio Success	Growth success	Opportunity success	Project success	Quality
Product renewal	0. 22	0. 24	0.29*				
Personnel recruitment		0.29*					
Cost reduction		0.43**					
Productivity growth	-0.28*						
Process innovation							-0.32*
Training		0.28*				-0.34*	
Protection intellectual assets		0.37**					
Technical co-operation				-0.39**			
Acquisition of the firm	0.31*				0.30*		

Notes

Strategic variables for the period 1988 to 1992 are listed in order of decreasing average priority (* denotes significance at the 0.05 level and ** denotes significance at the 0.01 level). Significant correlation coefficients in the range 0.3 - 0.4 are shown and remarkable correlation results.

Second, the overall success measure negatively correlates with productivity growth. Furthermore, a high prioritisation by companies for product renewal between 1988 till 1992 also made a difference to frontrunners' performances. This indicates that cost reduction and protection of intellectual assets drove frontrunners. The companies that regard the recruitment and retention of professionals as a high priority belong to the frontrunners. This is particularly relevant, given the fact that ICT personnel are a scarce resource (see chapter 7.6).

Differences in timing of company policy

The timing of policy items is important to innovation success. This is in line with earlier findings (Cobbenhagen 1999, Schreuder *et al.* 1991). Firms that address strategic issues at an early stage in their management agenda and sooner compared to other firms will be more successful. With respect to timing we found three key success factors:

- Successful innovators address more policy items and take these into account earlier than less success firms. To give an example,

ICT firms that have addressed cost reduction sooner have a better chance of being a successful innovator. The same applies to the recruitment of personnel and training. The sooner ICT firms address these strategic issues, the better their innovation process.

- Success requires long-term commitment and devotion to key policy items. To give an example, long-term devotion to product renewal is beneficial to the innovation portfolio of firms. This item is significantly correlated in both periods. The same holds true of long-term devotion to training and process.
- Successful innovators differentiate by addressing the following issues sooner: less emphasis on productivity growth, more emphasis on cost reduction, protection of intellectual assets and acquisition.

Looking at the reasons for timing differences, several case observations can be made about frontrunners:

- frontrunners encountered certain organisational issues earlier and hence had to address the policy implications early;
- frontrunners are more action-oriented and pro-active than less successful firms and are better at foreseeing policy issues by picking up weak signals from the environment and the organisation;
- frontrunners have a more integral perspective of strategy and innovation and therefore address issues broader and sooner;
- by the very nature of the company culture, frontrunners know when to change and get rid of built-in organisational routines and therefore concentrate on change issues earlier.

Let's further explore each of these arguments for ICT firms.

6.6 INNOVATION STRATEGY STYLE

The presentation of strategic priorities in the previous section indicates some of the strategic beacons and objectives of companies in the ICT sector. But what about the strategic discussions on innovation? Do managers discuss the innovation strategy as much as they discuss their business strategy? This section will present a further analysis to identify the strategic directions of ICT companies. In doing so, this study attempts to characterise how thoroughly a firm's strategy adheres to innovation in terms of the strategic focus on new products, markets and technologies. Figure 6.2

presents the innovation styles of the most successful innovators and the least performing companies. A correlation analysis (table 6.3) shows that five ratings are significantly correlated to overall innovation success. Although the outcomes should be considered carefully as the correlation ratings are not high, we can derive three interesting categories of outcomes.

Figure 6.2: Innovation strategy style

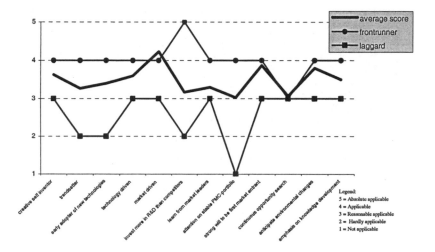

Frontrunners have an innovation style typified by:
- being a trend setter;
- investing relative more in R&D;
- desire to be first market entrant;
- early adopter of new technologies;
- invest in business opportunity searching.

Members of the pack seem to have two different styles:
- *Process driven*: exemplified by four following characteristics: early technology adoption, technology-driven strategy, anticipation of environmental changes and emphasis on knowledge development.

- *Growth driven*: typified by creative self-development, desire to be first market entrant and continuous opportunity searching.

Laggards have an innovation style especially characterised by being 'market - driven' and 'attention for stable product-market-customer portfolio'.

Table 6.3: Correlation results of strategic characteristics and performance

Innovation scorecard: Individual success items:	Overall success	Process success	Portfolio success	Growth success	Opportunity success	Project success	Quality
Trend-setter	0.37**						
Early adopter of new technologies	0.31*				-0.32*		
Technology-driven		0.34*					
Market-driven							
Invest more in R&D than competitors	0.24*				0.34**		
Creative self-developer				0.30*		-0.25*	
Strong desire to be first market entrant	0.35**		0.39**	0.31*			
Learning from market leaders		0.22					
Continuously searching new business opportunities	0.30*			0.25*	0.31*		
Anticipate environmental changes		0.24*	0.24*		0.27*		
Emphasis on knowledge development		0.38**				-0.24*	

Notes

(* denotes significance at the 0.05 level and ** denotes significance at the 0.01 level). Significant correlation coefficients in the range 0.3 - 0.4 are shown and remarkable correlation results.

First mover and trend-setter

The subject of first-mover has gained attention in the literature. Being a first mover reflects the drive and capability to be the first in launching new products or services. Porter (1992) argued that it illustrates the firm's objective to constantly broaden and upgrade the internal strengths in order to sustain and extend its competitive advantage. For the ICT sector, first mover indicates the firm's willingness to be at the edge of new products and services. A first-mover company tries to reap the economic benefits from entering a new business area. Being there first may result in colossal turnover or profit growth or a sustainable advantage.

> Company F is a medium-sized system developer and system maintenance company characterised by first-mover behaviour. The company has set up a separate unit for trend watching in the ICT sector. This unit, Exploration Informatics Institute, has several roles. The company wants to contribute to the domain of information science, where technical boundaries are disappearing rapidly. The task is to describe the new boundaries of informatics and train their customers on new technological possibilities. The other role for doing so is to inspire the client on new information systems. The unit consists of two full-time equivalents and a budget of 50,000 euros. These people travel around the world, looking for new groundbreaking developments from Silicon Valley, Israel or other high-tech development places. The unit publishes on trends such as the Internet, intra- and extranets, scripting technologies, systems development tools, etc. The aim is to be in the forefront of new technologies and try to create new business in the market. This philosophy renders them different from their competitors. But it also differentiates the company from their mother company, a large international ICT firm.

Hamel & Prahalad (1994) associate such company behaviour with companies that are willing to deal with risk taking, with new market creation and with 'expeditionary marketing'. Companies such as company F stretch their strategy. Kogut & Zander (1995) refer in this context to first movers as companies that have the strategic intent to develop unique and differentiating competencies that are hard to imitate or transfer.

Technology-driven
The analysis reveals that technology-driven firms that adopt technologies early excel in their innovation process. Several studies have advocated that technologically driven firms possess high levels of technical know-how and skills (Moeneart *et al.*, 1990; Booz Allen Hamilton, 1968; Allen, 1978). This study confirms that 'technology-driven' can contribute to innovation success.

Market-driven
The absence of a relation of the variable market-driven is remarkable. Market-driven reflects the quality of performing activities, such as preliminary market assessment, market studies and customer tests. Empirical studies have recorded this to be positively affecting innovation success (Cooper & Kleinschmidt, 1993). It might be argued that this is not a

differentiating factor between the ICT firms. All firms claim to be close to the market and interact with their customer. Just like the CEO of Company T:

> Company T is a software development company supplying administrative software packages. The company competes with a large European administrative software company. Company T is specialised in developing standard and custom-made software for the construction sector. It wants to build a long-term relationship with its clients, small- and medium-sized constructors. Their aim is to gain a good position in the segments of the construction market, which is concretised in the development and implementation of software. For instance, subject matter experts conduct software training for clients. These people with hands-on experience in construction are closer to the customer and know the construction sector inside out. This distinguishes us from our competitors... .

ICT services providers pointed out that customer demands drives renewal of their services, not technology.

Creative self-developer
Burns and Stalker (1961) already argued that a lower formalisation degree in an organisation contributes to a more open, receptive organisation, which is more creative and inventive. The origin of creativity can lie in different sources, e.g. a strong emphasis on searching new technology or markets in daily work practice, or from allowing slack resources in R&D. It can also be associated with the organisation culture of creative destruction, informal communication and dialogue (Abernathy & Clark, 1985). The creativity item in this study is significantly correlated with the growth success criteria. This might indicate that creativity is a rewarding attribute for companies in the ICT sector. In an organisation, creativity may substantially contribute to the uniqueness of new product designs and the innovative features (De Bono, 1994).

6.7 RESULTS: RESOURCE CAPACITY

The resource-based approach assumes that firms are a portfolio of resources, striving for market creation and competence building. To determine the degree of resources invested in innovation, we looked at several types of resources (see also Huizenga, 1997; Barney, 1991; Grant, 1991; Cooper & Kleinschmidt, 1995):

- R&D resources: R&D intensity, R&D allocation, R&D personnel;
- human resources: education and training expenditures;
- technical resources: technological position, portfolio of technologies.

A traditional resource variable in the literature is the investment of R&D resources, measured as a percentage of turnovers (Cooper & Kleinschmidt, 1995). This R&D intensity figure reflects the level of R&D investments made by a company. In the literature a debate is being conducted as to whether higher values of R&D intensity are related to innovation success (Rothwell, 1992, 1994; Freeman, 1988).

R&D resources
A correlation analysis on the R&D intensity variable, for successive time periods, revealed no clear pattern. Moreover, R&D intensity only seemed to have a little impact on the turnover growth resulting from new products.

R&D allocation
The relative amount of R&D investments only partly explains whether firms are successful at innovation. The allocation of R&D resources also matters. To what kind of innovation activities are R&D resources allocated, and are there differences in expenditure patterns? In order to identify possible differences in R&D spending managers were asked to breakdown their invested R&D resources into a predefined set of R&D activities. Figures in table 6.4 show that 60% of R&D went into product and service innovation and 35% into process innovation. This means that more than half of the R&D investments is focused on improving and renewing products and services. This finding might indicate that ICT companies are more inclined to invest in innovation through new products and services and less inclined to innovate through new or improved (service) processes.

Table 6.4: Allocation of R&D efforts

Percentage of R&D investments allocated to the following activities (n = 30):

R&D focused on product & service innovation	60%
R&D focused on process innovation	35%
Other	5%
Total	100%

Table 6.5 presents the statistical findings for R&D resource allocation. The analysis results indicate that R&D resources allocated to product innovation significantly correlates with both overall success (r=0.47*) and with project success (r=0.64**). The results suggest, first, that ICT firms spending more R&D on product innovation have a higher percentage of successful new product ideas. Second, ICT firms allocating more R&D to product innovation have a higher percentage of products and services actually being launched in the market.

R&D personnel
Another resource variable is concerned with the number of R&D employees (measured in full-time equivalents). This can indicate a possible association between the absolute number of people working on R&D and innovation success. In this study a split is made between:

- the internal R&D personnel;
- the R&D personnel, external to the company.

The latter can entail outsourcing of R&D activities, joint R&D projects or the hiring of external R&D professionals. When we take a look at the figures presented in table 6.5, the total amount of R&D efforts seems to contribute to portfolio success and growth success. A breakdown of the total number of R&D human resources reveals an interesting difference. Internal R&D especially contributes to the growth of the company, whereas the hiring of external R&D has a positive impact on the revenue renewal of the product and service portfolio of a company.

Human resources
Another question in the context of resource-based approach to innovation is how the success factors relate to companies' expenditures on personnel education and training. Especially the services part of the ICT sector is a knowledge- and people-intensive business. Investments in employees, through training and education, can be regarded, particularly for service providers, as an indication of the investments in innovation. Recent research indicates that successful service provides pay considerable attention to this subject. But is this valid for the ICT sector as well?

When we look at the results in table 6.5, the figures show no clear relation between the HR expenditures and the overall success factor. Remarkably, the HR variable is only correlated with the growth success factor for both time periods. One would expect the market to reward investments in human resources, especially if the labour market shows severe signs of labour supply shortage. One would expect firms that take good care of their employees by training and educating to be more successful. Yet the figures indicate no sign of reward other than growth of the firm, implying that company growth and increased personnel recruitment are closely related.

Table 6.5: Correlation of R&D resources and performance

Innovation scorecard: Individual success items:	Overall success	Process success	Portfolio success	Growth success	Opportunity success	Project success	Quality
R&D resources							
R&D intensity 1993-1994			0.42*				
R&D intensity 1994-1995			0.50**				
R&D intensity 1995-1996			0.55**				
R&D allocation							
product innovation	0.47*					0.64**	
R&D personnel							
Total # of R&D personnel			0.39*	0.35*			
internal R&D (in fte)				0.39*			
external R&D (in fte)			0.39*				
Technological resources							
technological position	0.31*	0.40**		0.37**			
portfolio technologies	0.31*	0.47**		0.39**			

Notes

(* denotes significance at the 0.05 level and ** denotes significance at the 0.01 level, results > 0.3 are shown).

Technological resources

The technological resources were measured by asking managers about:

- the technological position relative to their competitors;
- the technological portfolio relative to their competitors.

Table 6.5 shows a remarkable finding on the technological resources. The proficiency of technologies mastered by the firms correlates positively to overall innovation. We conclude that frontrunners appear to have a proficient technological position.

6.8 DISCUSSION: RESOURCE CAPACITY

R&D intensity
The study showed that the R&D intensity in the period 1991 to 1996 increased at a regular rate, both for software developers and service providers. In terms of high R&D spending, the software producers are the biggest spenders. Small software developers (less than 40 employees) invested more in R&D than large software producers, on average 36% versus 17%. The figures for large international software developers such as SAP, Oracle and Microsoft were 16%, 9% and 15%, respectively at that time (Booz-Allen & Hamilton, 1997). On average, the latter firms spend 26.5% of total turnover on research and development. In comparison, the large Dutch software developers (more than 250 employees) in this study spend a similar amount on R&D. Several reasons account for these differences in figures, for example the maturity of these firms, compared to the start-ups, differences due to the majority of revenues that might come form software products or from services. The maturity of the product life cycle is also an influencing factor. Some of the larger software producers face major new product releases, requiring extra R&D resources.

The average R&D figure of 4 to 5% for the service providers is in sharp contrast with these figures. But we also found a few exceptions of service providers that spend 10%, 14% and even 50% of their return on R&D. These service providers have spent R&D resources by building expertise centres and business development units for e.g. ERP software implementation, Oracle programming or client server migration.

Interestingly, R&D intensity only correlates with portfolio success, and not with any of the other success measures. Furthermore, this correlation is repetitive as it accounts for three successive periods. The result indicates that firms investing in R&D are continuously successful at developing profit- and revenue-generating new products and services.

Human resources
A remarkable finding is the absence of a correlation between training expenditures and innovation success. The ICT sector is perceived as a sector with a scarce supply of ICT personnel, while companies invest fairly large amounts in education and training. A detailed analysis of the education and training expenditures reveals the existence of large differences between firms. Expenditure rates differ from 0.5% of total turnover for some companies to over 10% of total turnover in education and training for

others. Differences in companies can be based on differences in opinion of holding on to employees to ensure a low percentage of personnel turnover. Additionally, companies seem to differ in opinion on employee training to ensure the value of an employee to the company. Why do companies spend over 10% on training? First, some companies with high expenditures do so to upgrade the skills of their employees, either because clients require up-to-date skills or because new technologies make existing skills obsolete rather quickly. These companies invest heavily in personnel to develop new skills: human resource management has a strategic goal (see cases below). Second, some companies show high scores on their education and training expenditures to strengthen their recruitment efforts. These firms try to differentiate in the labour market, e.g. through weekly recruitment campaigns and brand building in newspapers or special recruitment actions like flying with potential senior colleagues.

Training as starting capital

Company P developed out of an idea of university researchers and a lack of money. In co-operation with the local job centre a subsidised retraining programme was designed for unemployed academics. The training programme served as an investment in the development of a new software programming tool and the training of the people who would eventually sell the product and/or implement the software at the client's. The downside of this creative solution was the potential risk of a failed product. The market might not purchase the tool, and trained employees might end up with skills that are neither valuable nor requested by the market. The training programme succeeded. The first group of academics were the basis for a company that is known for its innovativeness in software programming tools and has started offices in other European countries as well as the United States.

Training as a core competence and new service

Company Q has developed as a market leader in the maintenance market over the years. It is known for its contribution to developing the IT maintenance market, based on ITIL standards. From the outset, the company recognised the importance of personnel policy and training as its driving force. The company developed this into a unique competence, which attracted a lot of potential IT personnel. Competitors have learned from this company's slogans in advertisements and its consistent broadcasting of their training philosophy. Company Q was able to grow in number of employees, no matter how tight the IT labour market was, by more than 20% annually. The approach used by the company had become even successful to the extent that it led to the creation of a new business line for company Q: 'human resource management for the IT sector'. Company Q sells the concept, methodology and techniques for a proper human resource investment programme to other companies both in the ICT sector and beyond.

Training investments in a shrinking situation: project Amadeus

Company G had a tough time during the years 1995 to 1998 in a flourishing ICT sector. Competitors received job order after job order, but the customers of company G were not interested in automation. The investments in ICT by the agricultural sector, company G's market segment, were extremely low. Scaling had a higher priority in the agriculture industry than automation. After the wave of take-overs extra attention was given to new automation, but not until early in 1998. In the context of this market development the question was what the company would do in the mean time, if customers were freezing their automation budgets.

Around 1995 to 1996 the consulting and engineering unit was composed of 70 employees. The employees knew the specialised 'tricks and trades' for software development. A total of 30 development environments were managed within the unit. Each employee could handle over five to ten software tools, but most of this knowledge was no longer up-to-date. An information system was built for each customer group, using different development technologies. The diversity of tools and systems resulted in a complex set of development processes to be managed. In 1996 company G decided to install four core development environments and databases: AS400, Oracle, Progress and Delphi.

In order to monitor the changes, the company launched an education trajectory under the name 'Amadeus'. Its aim was to train the engineering specialists in concert in one single software tool. At the same time, a reorganisation of the consulting and engineering unit started. The total budget was an additional sum of €350000 on top of the existing training budget. Remarkable decisions especially considering the situation revenues were dropping. The company has reserved a substantial amount of its profits to invest in knowledge improvement. The trajectory was planned to take place over a period of two years and was to be implemented later with the unit system control.

The cases illustrate the deeper and complex situations of human resource management in ICT firms. The motives for human resource investments differ between firms, considering creative solutions such as subsidised training in company P, marketing HRM as core asset in company Q, or even taking decisions as in company G to reserve a substantial amount of the profits for knowledge investment.

Technological position

Both the technological position and the portfolio of technologies are key success factors for ICT firms. They tell us something about the absorptive capacity of ICT firms. Absorptive capacity refers, for example, to the extent to which new technologies are identified and incorporated. The study found contradicting evidence on the technological position and portfolio of technologies. The ICT study found evidence that the ICT firm's technological absorptive capacity influences a firm's innovative success. In particular the extent to which the ICT firm is prompt to detect and adopt technological developments. Just like company L, which is a software developer specialised in administration software for governmental institutes:

> The company has installed a team structure to give a continuous impulse to innovation. Domain teams have been set up that represent both the technology and marketing perspective in ICT projects. Company L has installed two development teams, one team with the task to explore and adopt new technologies in the environmental sector, and the second for the development of new technical systems. In order to ensure the process innovation one overall team has been set up. The goal was to stay at the forefront of new facilities, tools and system technology and exploit them in the ICT projects.

The use of these teams helped the company to absorb new technologies and tools. This approach would help in building more flexible software and hardware solutions. The AS400 platform was the core system for company L. Because of the teams the company could now expand its systems to HP9000 and IBM RS/6000 platforms and absorb the technology behind the systems. In this way the company could handle the client's requests for open systems and customised software packages.

7. Organisation, Process and Innovation Performance

7.1 ORGANISATION STRUCTURE AND TEAMS

Managers were asked to typify the organisation structure of innovation processes. Based on a typology of five basic organisational structures, we were able to classify the companies' organisation structures (Cobbenhagen, 1999; Larson & Gobeli, 1988; Galbraith, 1973). Table 7.1 describes the respective organisational characteristics and reports on the ICT firm response, ranging from the pure functional organisation reflecting a functional specialisation of innovation to the project team-based organisation structure. Frontrunners prefer a project based or project matrix organisation.

The organisation literature argues strongly in favour of the use of multidisciplinary teams to innovate (e.g. Takeuchi & Nonaka, 1986). Multidisciplinary teams organise around the integral development task itself, rather than focusing on the single contributions to the overall product outcome. Managers were asked to describe their use of teams.

Analysis was conducted to see if both the organisation structure and the multidisciplinary team affect innovation success. In line with earlier empirical findings (Nonaka, 1990; Cobbenhagen, 1999; Larson & Gobelli, 1988) the organisation structure was found to be a key success factor in the ICT study ($r=0.32*$). However, the multidisciplinary team variable was not significantly correlated with any of the success factors ($r=0.17$).

The ICT firm responses indicate that firms with a functional orientation are in the minority. Around 75% of the ICT firms have organised innovation through a matrix, project matrix or project team structure. One-third of the ICT firms have opted for the project team structure compared to 18% of the frontrunners in the preceding study. The question in the present study is whether the extent to which the firm's innovation processes are organised horizontally (project-wise) differentiates the successful companies from less successful companies.

Table 7.1: Organisation of innovation process

(1) % of ICT firms (*) (2) % frontrunners cross-sector study (**)	(1)	(2)
Functional organisation An innovation project is divided into segments and assigned to relevant functional (sub)departments. Departments conduct project tasks successively. The project is co-ordinated by the functional managers and higher management echelons.	12	3
Functional matrix A project leader with limited authority is appointed to co-ordinate the project across the respective functional (sub)departments. The functional managers have responsibility for and authority over their specific segment of the project.	12	24
Matrix organisation A project leader is assigned to oversee the project and shares responsibilities and authorities with respect to the completion of the project with the functional managers. Decisions are made jointly by the functional managers and the project leader.	24	15
Project matrix A project leader is assigned to supervise the project and has the primary responsibility and authority for its completion. The functional managers provide human resources and the necessary functional knowledge and capabilities.	20	40
Project team organisation A project leader is put in charge of a project team composed of a core group of people from all relevant functional disciplines. They are allocated to the project on a full-time basis. Functional managers do not have any formal involvement.	32	18

Notes
(*) ICT firms responses based on 120 interviews in 32 firms.
(**) % distribution of among group of 62 innovative frontrunners and members of the pack, based on cross sector study (Cobbenhagen, 1999).

Multidisciplinary team approach: some empirical findings
Ancona & Caldwell (1990, 1992) examined 45 product development teams in 5 companies on their success patterns and the ways of external communication. Success was measured by subjective team and management ratings of performance. The study observed that team members communicate more with outsiders who had similar functional backgrounds.

Thus, when multiple functions were represented on the team, there was more external communication by the team as a whole. The issue of internal communication has been explored extensively (e.g. Dougherty, 1990, 1992; Dougherty & Corse, 1995; Imai, Nonaka & Takeuchi, 1985). Keller (1986) found that internal group cohesion helped improve performance. Ancona & Caldwell (1992) found similar evidence on the role of internal communication among team members. They observed a contributing role to performance by strong internal communication on goal definitions, development of work plans and prioritisation. As suggested by Dougherty (1990, 1992) multidisciplinary teams can overcome cross-functional communication barriers. Especially when team members participate in concrete tasks together and violate routines such as division of tasks. Multidisciplinary teams are known to break away from built-in rigidity and daily routine practice, which can hamper change (Leonard-Barton, 1995).

It is remarkable to find that ICT firms that organise their innovation projects with multidisciplinary project teams do not appear to be more successful innovators than firms using a functionally oriented approach. The correlation analysis results do not confirm earlier findings on the importance of a flow-oriented and multidisciplinary method of organising innovation (Cobbenhagen, 1999; Cooper 1994; Takeuchi & Nonaka, 1986). The clue seems to be that up to 80% of all ICT companies work with a core team that is facilitated in each of the sequential project phases by temporary team members. For the majority of the sector it is common practice.

Managers advocate (e.g. Barczak 1995; Stopford & Baden-Fuller, 1994) a team approach and argue that the 'entrepreneurship' conditions allow for a more effective application of scarce resources within shorter time schedules. In the ICT sector the interviewed managers stressed the importance that without teamwork it is impossible to secure the cross-functional interfaces, which are essential to the business services sector. Cooper *et al.* (1995) are among the numerous authors (e.g. Galbraith, 1994; Den Hertog & Huizenga, 1997, 2000) who argue that multidisciplinary teams can help improve decision making and implementation of new services. Especially where barriers have to be taken. Just like company F.

Company F is a small software company and market leader in the area of postal management software. They introduced 'skunk work' teams. The company employs 30 people. Three years ago a new development project was started, focusing on the development of a new release of an existing software package. The new product was to feature a graphical user interface (GUI). Object-Oriented

(OO) technology was to be used as design principle. However, the development process seemed endless. The developers failed to develop a GUI-based software package, and the project started to drift. While looking for reasons for the project delay, the CEO found out that the skills composition of the project group was not suitable for the project goal. The group members were programmers that used second- and third-generation programming principles, which were drastically different from OO. In the project group, the programmers used the product structure of the old product. They kept on reprogramming the old product, without making real progress towards a GUI-based product. The management team chose for a radical solution. The best project leader was set free from regular business operations. Together with four new graduates from a technical school and an expert on user ergonomics he became responsible for the project. This project team was set apart from the regular organisation and located in a separate building, with full responsibility and their own budget. A process facilitator was incorporated to accelerate the teambuilding and team development. The new post management software product was released within six months without exceeding the budget. This company is now the largest supplier of the product in The Benelux.

In *'A Passion for Excellence'*, Tom Peters (1988) promoted skunk works as highly innovative, fast-moving, and slightly eccentric activities operating at the edges of the corporate work. The origins of 'skunk works' go back to a development team at Lockhead's Design and Development Centre that developed radical new product concepts by using fully independent heavyweight project teams.

7.2 PROJECT MANAGEMENT

In order to explore the relevance of project management in the ICT sector we have tried to measure the presence and quality of project management and its impact on innovation success. In doing so, we have mainly focused on:

- *project planning* dimensions, as a measure for the degree of detailed planning of an innovation project;
- *project responsibility* dimensions, as a measure for the degree of responsibility and accountability of the project leader and the project team.

Both dimensions are an attempt to capture the extent to which companies have emphasised the role of up-front development work and detailed market and technical assessment. Figure 7.1 provides the average figures for project planning. The figures show that companies' plans start from financial targets (profit and turnover) and go/no go decision points. Remarkably, less emphasis is placed on planning costs and time schedules.

Figure 7.1. Project planning

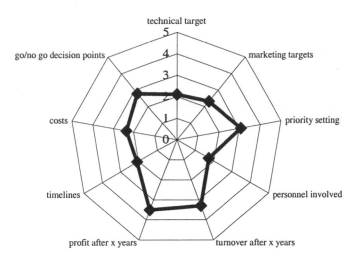

Notes
Score:
1 = not mentioned in innovation planning
3 = roughly mentioned in innovation planning
5 = concrete defined in innovation planning.

Project planning
In accordance with other studies (e.g. Cooper & Kleinschmidt, 1995) the following planning items contribute to innovation success:

- clear priority setting (r=0.26* with overall innovation success);
- making the profit and turnover goals explicit up-front (r=0.26* with opportunity success);
- having clear go/no go decision milestones (r=0.29* with quality). Clear milestones add to the quality, particularly for a service environment.

This supports work on the stage-gate model of innovation (Cooper, 1994; Ancona & Caldwell, 1992, Eisenhardt & Tabrizi, 1995), which emphasises the explicit use of go/no go decision points.

Project responsibilities
The overall question is whether project leaders should attain broader and deeper responsibilities, or companies should use cross-functional teams. We have observed that responsibility for the technical design (r=0.40* with growth success), project alignment (r=0.32* with portfolio success), project cost and team composition (r=0.39* with opportunity success) each correlate with innovation success. Moreover, companies that make project leaders responsible for cost control and team composition better seize new business opportunities. This responsibility contributes to the realisation of profit growth and the opening of new markets. This is a new finding seldom tested in the empirical literature. We have also observed that companies spend time to professionalise their project management, while at the same time supporting the use of teams. Just like Company D's method of project life cycle management.

Company D is a multi-vendor service provider and part of an international ICT service company. Three types of teams are used in this organisation: project teams, on-site teams, and remedial teams. The project teams are the most important for company D. They are responsible for the pre-sales, sales and after-sales cycle with the client, including contract preparation, contract implementation and actual service support. The project teams are steered by team leaders with clearly defined responsibilities. The project planning and project management are formally described in the Multi Customer Solution Project Life cycle. Each client process is co-ordinated along the defined project life cycle, whose main phases are as follows:

    1. The opportunity creation phase
    2. The contract preparation phase, including:
       - opportunity evaluation
       - development and proposal
       - negotiation,
    3. The contract implementation phase, including:
       - implementation
       - operation (support)

This project approach has been applied in the company for two years. A central role in this approach is given to the Project Office, which has to ensure that the formal management guidelines are actually followed and the project procedures, customer contacts and lost deals are written down. For each project it must be clear who is responsible for what and how the project is organised. The Project Office also has the responsibility to ensure proper reviewing at several stages in the project phases, such as an opportunity review or a bid investment approval review. The reviews have a multidisciplinary character. In addition to the core project team, the sales department, the business manager and the service delivery department are also involved in the reviews.

About half the companies reported to invest in new project workflow methods and supportive information systems, all of them aimed at speeding up and tightening the pre-development work and the design of We have come across a large number of projects using Rapid Application Development programmes (RAD). This suggest that the professionalisation of IT project management was a serious issue at the time of the company visits. A number of remarkable statements concern the role of programming. Research on passing cross-functional borders in the industry (Den Hertog & Huizenga, 2000; Roussel, Saad & Erickson, 1991) argue that programming of development work will be effective only if the organisation has streamlined its development process. In particular, it includes the introduction of formal control systems such as PERT (Project Evaluation and Review Technique) and CPM (Critical Path Method) in up-front work. The managers suggested that task allocation, work standardisation and formalisation are beneficial in programming innovation.

7.3 ORGANISATION CULTURE

A thriving positive climate for innovation is not determined by structure and project management systems only, but also by the values and norms prevailing in the organisation. These 'soft' organisation aspects are referred to as the culture of the organisation. Culture is often addressed as a success factor (Pfeffer, 1981; Moss Kanter, 1983; Johne & Snelson, 1988). We explored a number of cultural values and norms without explicitly conducting an in-depth culture survey. Scott Morgan (1994) refers to 'the unwritten rules of the game' in an organisation. In order to explore the culture for innovation managers were presented 9 culture variables. We asked them how they perceived the company culture, by asking 'describe the way things are going around here' (e.g. Senge, 1990).

Table 7.2: Correlation coefficients of culture dimensions and innovation performance

Innovation scorecard: Individual success items:	Overall success	Process success	Portfolio success	Growth success	Opportunity success	Project success	Quality
Hierarchical	-0.27*			-0.28*			
Conservative	-0.33*			-0.25*			
Competitive	0.24*						
Structured			-0.26*				
Uncertainty-avoiding					-0.30*		
Result-oriented							
Formal							
Process-oriented			-0.25*				
Open to change							

Notes

(* denotes significance at the 0.05 level and ** denotes significance at the 0.01 level). Significant results are shown.

No surprising results have appeared with respect to culture (see table 7.2). It seems to be apparent from the analysis that hierarchy and conservatism are cultural values that characterise the less successful innovators. These values also seem to be in conflict with team functioning. The successfully innovating companies seem to be characterised by an internal competitive culture. Structuralism, uncertainty avoidance and process focus are cultural values that partially differentiate the laggards from the frontrunners. Stopford & Baden-Fuller (1994) already argued that most organisation scholars regard bureaucracy as the antithesis of entrepreneurship [and innovation].

Findings support Vos' (1985) and other's (e.g. Quinn, 1985; Damanpour, 1991) opinion that a risk-taking climate is beneficial in development projects in the software industry. Based on an in-depth case study of Microsoft, Cusumano & Selby (1995, p. 416), indicated that innovation were present through continuous self-critiquing, feedback and sharing:

> The open culture of Microsoft is still not too far away from the loosely organised world of hacker programmers. Its people abhor political 'turf battles' as well as

bureaucratic rules and procedures, unnecessary documents, or overly formalised modes of communication. As a result, individuals and teams act quickly on issues they feel are important.

7.4 KNOWLEDGE DEVELOPMENT: INTERNAL AND EXTERNAL CO-OPERATION

Our research has focused both on company-internal and company-external issues for new business development. In the preceding sections, support was found for matrix structures being the most preferred structures to allow for cross-lateral work. The subject of cross-lateral relations can be extended beyond the enterprise. Below we will address the knowledge relations that ICT firms have established and their relevance to innovation. We will examine (Den Hertog & Huizenga, 2000):

- knowledge development resulting from external and internal co-operation;
- the degree of external co-operation;
- the types of knowledge co-operation in the innovation process;
- knowledge relationships with (potential) customers and suppliers.

Knowledge relations can be of strategic importance (Huizenga, 1997, 2002b), e.g. for the joint application of expertise in development areas and the sharing of research and/or development costs. Just like company S.

S is an international software developer of ERP systems, which faced a lack of knowledge of their end-customer's business. Despite the fact that their service partners (large implementation specialists) were implementing software, the company noticed that this customer knowledge was crucial to the development of new software. Based on the knowledge they needed and the markets they wanted to serve. The company therefore developed the Enterprise Modeller. They asked their service partners to jointly develop a reference model, incorporating business process knowledge for each market. The aim of the Enterprise Modeller was to acquire a process view of the organisation as a means of configuring the software. The deal included the exclusive right to use the reference model for the service partner. Company S was to be the owner of the model and to possess the necessary customer knowledge, while its partners could build up expertise. Reference models were developed for the automotive, electronics, aerospace, and chemicals industries.

We asked managers about the total efforts they had made in knowledge development. To obtain some indication, we used a somewhat new procedure. As an experiment, we supposed that it was possible to weigh knowledge in terms of relative value (e.g., money, full-time equivalents). For example, €100000 of knowledge development activities might be linked to internal efforts. With this assumptions we might be able to recover the wellsprings of knowledge (Leonard-Barton, 1995; Den Hertog & Huizenga, 2000; Huizenga, 1997). The procedure showed a number of interesting outcomes.

The question was stated as follows: 'If the total knowledge development efforts of your company are set equal to 100%, how would you allocate the company's total efforts in knowledge development originating from internal and external sources?' The figures (table 7.3) show that on average 59% of all knowledge development efforts are based on external sources. Internal knowledge development efforts account for 40% of total efforts. When subdividing external knowledge development, the figures indicate that around one third of the knowledge created is derived from co-operation with customers. Suppliers account for 16% of external knowledge creation. Knowledge institutions, in particular universities and related research institutes, account for 5% of external knowledge development. Knowledge development through direct acquisitions only matters for a small set of firms, and is responsible on average for 3%.

Table 7.3: Distribution of knowledge development activities

	Total % of knowledge developed	Standard deviation
Internally developed knowledge	41%	(22.1)
Externally developed knowledge:		
• Co-operation with customers	28%	17.1
• Co-operation with suppliers	16%	15
• Co-operation with university/research institute	5%	4
• Acquisition	3%	6
• Other sources	7%	11.4
Total knowledge development efforts	100%	

First observations show that:

- External sources are more often used to develop new knowledge, and that in particular customers and suppliers are the major external sources of knowledge development;
- In this knowledge-intensive sector, firms seldom co-operate with universities. Yet we have observed that those few firms that acquire knowledge developed in co-operation with universities score highly on overall innovation success. The variable correlates positively significantly (r=0.45*) with overall success;
- Firms that develop knowledge through co-operation with customers score highly in terms of growth. The variable correlates (r=0.39*) positively significantly with growth success;
- A few firms gained new knowledge through external acquisitions. These firms have a high success score due to their ability of better seizing new opportunities and realising new profitable products. The corresponding variable correlates (r=0.38*) positively significantly with opportunity success;
- Knowledge developed in co-operation with customers accounts for 30% of all knowledge activities.

Although the correlation coefficients are not high, the outcomes are interesting. No significant correlation was found between any of the success factors and the internal knowledge development practices. This might imply equal scores for frontrunners and laggards. Alternatively, it might imply that external knowledge development contributes more to innovation success than internal practices. Cohen and Levinthal (1993) and Szulanski (1996) already argued that the 'absorptive capacity' of an organisation is of importance to innovate. Just like the absorptive capacity of frontrunner company A.

Company A is a separate business unit of an international ICT firm. According to the CEO the business unit is the major innovator for all the European practices. It spends 50% of its turnover on R&D. Most R&D activities are directed at scanning new technologies and potential markets and application areas. They aim to pick up early new technologies, such as datawarehousing and HTML web development, and transforming them into services that substantially contribute to the competitive competence of clients. Company A views *innovation as a race at the assembly line*, where new ideas come by continuously and at an increasing pace. The power of innovation is to scan the assembly line and pick out the most promising trends. Together with a group of early adopters of trends, expertise on pilot projects is developed. At the same time insight is gathered into the market potential of the product or service.

The results confirm the argument that exploring and exploiting external information and knowledge sources is conducive to innovation. Company S explains their knowledge philosophy and absorptive capacity.

Company S is a large software and service enterprise consisting of several market divisions. It is organised along a divisional structure and has a separate innovation unit. The Chief Innovation Officer (CIO) of the unit states: '… the unit has the ambition to be the architect of the software factory …'. We are developing a vision of streamlining the software process similar to a production process, to become the standard for all divisions. He goes on: '… our software engineering philosophy has to leave the pioneering phase …'. Most software projects are custom-made projects, but the unit wants to introduce a new software development paradigm. It monitors the industrialisation process of software development and looks at the relevant developments. Their ambition is based on their knowledge of:

- Software development methodologies such as joint application development (JAD), rapid application development (RAD) and object-oriented programming (OO) and the 'traditional' stepwise software development methodology (SDM);
- The organisation of software engineering;
- Insight into development tools and development environments.

The CIO explains: '… New development paradigms have to come from radical changes in the current way of working. Our divisions cannot rely on an evolutionary development approach. We need to introduce a paradigm shift in the software production process, otherwise it will not have any effect. Paradigms like OO, rule-based engines and component thinking all entail a different context. What we need are paradigms well beyond the traditional 'waterfall-like' SDM method of software programming. There is a different investment philosophy behind them, and steering models like function point analysis no longer work …'. The CIO said that several pilot projects on RAD and JAD were to be started in the divisions within six months to absorb new knowledge.

This outcomes are surprising for a knowledge-intensive sector of which we would expect more co-operation, especially because a large number of ICT start-ups originate from universities and polytechnic schools. Only a few firms differentiate by developing knowledge through contacts with universities and academic research institutes. Interview data revealed that such co-operation could take different forms, such as joint research and development on ICT issues, as well as graduate programmes with universities to recruit new employees.

A last observation is concerned with the role of suppliers. Suppliers as a source of knowledge do not affect the innovation success measures. Managers explain this by pointing to the use of 'proven' technologies by ICT firms. Most companies use the same technology standards for software development or e.g. ITIL standards for infrastructure maintenance. These standards are predominately set by US companies. Also, at the time of the study the ICT firms were hardly involved in technological co-operation agreements with US ICT firms.

7.5 THE ROLE OF THE CUSTOMERS AND SUPPLIERS

This study has demonstrated that the market is a source of inspiration for innovation. Ideas may originate from new supplier products, a direct request from a customer, or an experience of an (un)satisfied customer. But the role of the customer is sometimes a drawback. Just like company Q experienced.

Company Q is a small software company specialised in wage and salary administration. Its market consists of small- and medium-sized accountant and administration offices, which use the software products for up-to-date fiscal and salary administration. The priority of the development unit and the CEO is not to acquire new market segments. The quality of current software products is improved to meet new technical standards, and to a lesser extent, to customer requirements. During the interview, the CEO admits that the customer group has one disadvantage. They are rather conservative to renewal and especially to adopting new products and services. This hampers the company's business development. In the long run, the company would like to broaden to other markets based on its core assets: fiscal and wage administration know-how. New services and the development of a personnel information system are the new options under investigation. However, the management team of Company Q does not know if the company can release from the conservative influence of its customers.

Most innovation studies argue for a close involvement of customers. Customers can co-create value. The question we raised in the survey is whether close co-operation with customer and/or suppliers is actually beneficial to innovation.

The first results are provocative. They show a strong negative correlation (r=-0.55**) between customer involvement and the renewal of service and product portfolio, indicating that too much involvement of these external parties is not beneficial to success. The empirical literature (Thomke & Von Hippel, 2002; Von Hippel, 1986; Cooper & Kleinschmidt, 1995) has revealed the importance of close co-operation with customers to create value. The managers interviewed in the ICT study also emphasised the importance of this subject. However, the case studies seem to be in contrast to these expectations. Cobbenhagen (1999) had also found this result.

Conservative or advanced customers?

One reason for this remarkable finding might be that the customer interaction is much more complex in reality. The business development manager of an specialised agricultural software company explains why innovating is complex.

> The company's customers range from high-tech frontrunners to traditional agri-food customers. He observes that the company needs to have solid knowledge of the business processes, the agri-food value chain structure as well as the customer needs in agriculture. The information flows in the production chain are the core business for the company. The company has difficulty in keeping pace with some of its customers. Large flower and fruit auctions in The Netherlands are frontrunners in using new technologies for a more efficient auctioning process. The company must go along with and even stay ahead of these frontrunners in e-commerce developments, e.g. to trade products via the Internet. If it fails to do so, the company is out of business as a supplier for these advanced ICT users.

At the same time, this company has a large customer base of small agriculture organisations that invest little in automation. The company used to develop a separate software package for each customer. The Engineering & Consulting unit employed 70 people and was equipped with over 30 software development tools to make the customised software. Each job was customised. The business development manager remarked:

> It is crucial to keep the technologies up to date, but that would cost too much. Therefore, we decided to specialise on a few core technologies for our core groups of customers.

Customer interaction

Customer interaction and that renewal in the ICT services sector is almost by definition in a paradox. A more fundamental paradox for ICT companies, which we can formulate as: 'The ICT company must do what the customer tells' versus 'The ICT company must tell the customer what he should do'. The case example of company Q shows how the role of the customer was used as an argument for internal politics to carry on as before. In the agriculture software firm we observe the opposite. The customers expect the ICT service provider to be ahead of them in terms of supplying new market and technology knowledge. The firms seem to be tightly linked to their customers and depend on them for new product development. We might speak of a 'path-dependent' relationship between the company and its customers (Huizenga, 2002a, 2002b).

This kind of customer interaction in the service sector is less investigated in the innovation literature. The examples reveal a mutual interaction, which seems to be rooted in a customer-driven knowledge development process, where ICT companies should provide new knowledge to their customers, e.g., about the opportunities of selling through the Internet. At the same time, they should depend upon them to build up customer and product knowledge. Various arguments can be given why frontrunners seem to better handle this interaction, for example:

- market pull does not seem to dominate in frontrunner companies but is balanced by a technology push;
- frontrunners set strict limits on the customisation of services and software;
- frontrunners limit the degree of modification of new products;
- or frontrunners let customers pay for special orders, extra software functionality and unique services specifications.

Company I even made a radical move to set strict limits for customisation by abolishing their customer group.

> Company I is a software developer that had suffered losses for more than three years. Its software development unit is in close contact with its clients. A selected group of clients, semi-governmental organisations, even has a special status including involvement in an official user group. This group provides monthly input on products they might need or would like to use. The user group discusses the extensions of software functionality, ideas on new applications and is involved in test scenarios. The interim CEO put an end to the special status of the

user group immediately after his entry. He remarked: '... Although the user group provides input for the software development unit, the members of the group are not obliged to buy any of the new products, product extensions or accompanying services. Due to these intense customer contacts, our software developers have made products that did not generate any additional revenues... a waste of our valuable efforts'.

More research to investigate the need for tight upstream and downstream customer relations in the innovation process can shed new perspectives on the role of the customer in knowledge and service intensive sectors.

Customer co-operation

We asked managers about the co-operation with customers during the several stages of innovation. Table 7.4 presents the statistical findings.

Table 7.4: Correlation results of external co-operation variables and performance

Innovation scorecard: Individual success items:	Overall success	Process success	Portfolio success	Growth success	Opportunity success	Project success	Quality
Customer involvement in:							
Ideas for process improvement		0.40*					0.39**
Implementation of process improvements		0.32*					0.43**
Service/ product idea generation		0.31*					0.27**
Service/product testing							0.28*

Notes

(* denotes significance at the 0.05 level and ** denotes significance at the 0.01 level). Significant results are shown).

Results

Customer involvement is relevant to frontrunners in two areas:

- improvement of quality;
- process enhancement.

Customer involvement benefits at several stages in new process improvements and new product ideas and product testing. This implies that customers can help streamlining processes and generate new service and product ideas. This finding suggests that frontrunners differentiate by involving customers in these particular areas. When we move on to the qualitative stories, we come across more insights.

Company T is a small software developer specialised in document management systems and workflow management systems. With 250 customers, the company is a niche player but market leader in postal management systems. Its customers include service and public sector companies with high correspondence traffic, e.g., chambers of commerce, universities, the police, and public utility companies. Innovation in company T involves a great deal of interactions with customers. The first activity is to identify all the customer needs. After that, a 'specification team' is set up composed of three software developers and three members from the user organisation. The 'users' are selected on the basis of their ICT knowledge and customer record behaviour. The composition of this team reflects the emphasis placed on domain knowledge above automation knowledge. The development process is not structured and there is no pre-determined development tool environment. The only structuring aspect in this stage of innovation is the use of prototyping. The software functionality is designed on the basis of the expectations and requirements residing in the specification team. Apart from the specification team, a sounding board group is set up, consisting of members from the user organisation, which assesses whether the specifications live up to their expectations. The CEO defines the drawbacks of this approach as follows:

- too much customer orientation does not deliver saleable new products;
- too much customer orientation in the design and development stage produces a large amount of unused software functionality.

The CEO points to the difference between the 'users' and the 'buyers'. The 'users', those who apply the software in daily practice, look at software dependent on the tasks they will conduct. The 'buyers', those who decide on the actual purchase of the software, are more trend-oriented and look at new ICT features like flashy graphical user interfaces. In the product development stage the role of the user organisation is heavily represented. The CEO remarks that neither this nor the presence of a sounding board group assessing the specifications however guarantees that the software package will imply a selling record.

The CEO is certain that innovation for his company is not by offering new functionality. In his opinion, innovation should be effected from a vision of mass customisation, where the customer configures the product, while the software producer delivers the product through a flexible production process. Software packages then become custom-made, on the basis of standard modules that can be produced in a standard process. The challenge is to find a balance between the diversity in customer needs and the need to simplify the production process.

Management learnings
What are the management learnings on the role of the customer? The motto of customer involvement seems to be contradicting. On the one hand, ICT companies claim to be customer-driven, and that it is important to listen to what customers have to say and need. ICT companies preach a client-oriented innovation process and are aware of the value of marketing as a source for new product developments. In reality, however, customer orientation seems not to be a factor that discriminates the frontrunners from the ICT companies lagging behind. But is this actually true? We have observed that customer information is beneficial, but at the same time that companies differ in the extent to which customers are recognised and used as a booster for innovation.

Customer paradox: learning from customers?
There is a customer paradox in service innovation, where companies need to balance between 'the company can learn from what the customer tells' versus 'the company must tell the customer what to do'. There are several explanations for this 'customer learning paradox'. There is a difference in learning from '(potential) customer panels' or 'getting feedback from current users'. But also the paradox inherent in the thin line between 'nice-to-have' customer needs and 'customer requirements' might lead to excess product and services functionality. Furthermore, there appears to be a difference in customer perception versus action. Similar to Argyris and Schön (1978), who distinguish between espoused theories and theories in action, there seems a difference between what customers and managers really think (espousing) and how they act (acting). Espoused theories represent the official innovation style, while the theory in action represents how managers actually act. Similarly, we got the impression that the managers think they listen to the customer and are driven by the market, but in reality act different in terms of involving the customer. The frontrunners however give customer involvement a different meaning (Huizenga & Veldhoen, 2003). Their leaders participate in customer panels, they discuss step by step the customer process and collect complaints. They do not focus on customer satisfaction, but deepen into the unaddressed and hidden needs of customers and try to provide solutions to customer problems.

7.6 HUMAN RESOURCE MANAGEMENT

In chapter 6.5 we addressed the strategic priority for personnel policy and human resource development. This study also confirms that training and education expenditures contribute to innovation performance (see also Becker & Gerhardt, 1996; Schuler & Jackson, 1996; Van Sluijs *et al.,* 1991). We already touched upon the relation between company growth and education and training resources. Two lines of reasoning are valid. First, firms invest in training because they achieve high growth rates and training ensures the up-to-date availability of knowledge and skilled people. Second, firms realise high growth rates *because* they invest in highly educated and trained employees. Both lines of reasoning might be true for the ICT sector.

This section presents additional analysis on the practices of human resource management (HRM). First of all it is remarkable that we did not observe any interesting outcomes on HRM differences between the firms. Especially remarkable for a 'people business' like the ICT sector. The majority of the ICT companies reported to conduct relevant activities on each of the HRM subjects. In order to measure the organisation's HRM practices, the research addressed four dimensions of personnel policy. The survey results all point in the same direction: HRM is an important subject, but does not differentiate frontrunners in innovation in the ICT.

The presence of a formal education plan for employees
Over 80% of the firms reported to have a formal education plan for their employees. No relation was found between one of the success factors and having an 'education plan'. This is not surprising, since for the majority of the companies in the ICT sector it is common practice. This suggests that an education plan does not differentiate frontrunners.

The planning horizon of an education plan for individual employees
The average planning horizon of an education plan averaged between two and three years. No significant differences were was found between firms. Again, for the majority of the companies in the ICT sector it is common practice.

The planning of career paths
The survey reported the same planning horizon with respect to career planning, i.e. two to three years.

The nature of individual performance appraisal
The survey results indicate that 90% of the firms conduct formal appraisal interviews with employees once a year. About 70% of the firms state that they conduct a formal career development interview once a year, so it is not surprising that the figures show no signs that frontrunners perform different. However, the question is whether these outcomes are correct, shouldn't we investigate other possible factors that play a role? The next case examples show the actual differences in the 'people business'.

> Frontrunner P posts ICT infrastructure maintenance and change management personnel with their customers in such industries as banking, insurance, chemicals, government and construction. In recent years, its growth figures have risen by 30%. The company uses environment scenarios to determine the company growth policy. The dynamic labour market and the dynamic technological development are the pillars for its growth strategy. In case the technology crystallises and the labour market stabilises, this has a major growth impact on the company. Also, if the labour market remains tight and the dynamics of the technology remain high, this influences the growth policy.

The CEO points out that the actual truth is somewhere between the two, but that their education policy should incorporate these impacts. For example, in one year over 3,500 applications were received for only 300 jobs. The recruitment and selection procedure is strict. According to the philosophy of the company, employees must have the capability to reach the top. A detailed training programme by which new employees are posted with different customers, complemented with successive training periods, reflects this philosophy. Career planning is oriented towards individual growth opportunities. The company measures the average progress of employees. On average 6% of the employees leave the company after one to two years. For 2% of the employees company P agreed that the career perspectives were better outside company P. The company received an innovation award for its progressive HRM policy in the ICT sector.

In comparison company H is a company posting ICT personnel. The CEO has been with the company for two months. He is a 'people manager' and wants to catch up on training, due to the strong penny-wise attitude towards training by the former management team. Company H's view on personnel is that training will result in more knowledgeable people. The CEO is clear on one thing: more knowledgeable people can generate more revenues for the company. In this, he differs from the pervious managers, who aimed at profit maximisation at the lowest possible costs. The CEO remarks:

Through training, our employees can be posted at higher positions and obtain consulting fees at the customer side, so they will be program managers and software project leaders instead of posting software programmers and software maintenance employees.

The CEO started career planning interviews with the employees. The CEO soon found out that the best people had already left the organisation and the remaining employees were less motivated. The ICT sector is a 'people business'. The people represent the invaluable knowledge needed to deliver services and make products work. Taking care of these people is the distinguished task of ICT companies. But companies differ in their motives.

8. What Differentiates the Frontrunners?

8.1 INNOVATION SUCCESS: OVERVIEW

The analysis of ICT firms and their practices and performance has yielded a large set of success factors. Some of them had been expected, based on recent research insights. A number of outcomes that are considered to be crucial in the literature are noticeable for their absence in this study. For example the role of customer involvement, multidisciplinary teams, and human resource management. These items do not contribute to success. The cases have provided insights why this happens. Other outcomes are more provocative in that they reveal new areas of management attention for ICT companies. We materialized new theoretical constructs like resource based ideas, path dependency, the role of technological stock and flows of resources. In order to assess the relative importance of the findings of the study, this chapter will summarise the results and give directions for innovation management and research. It tries to provide the followers with suggestions on how to escape from their position and join the frontrunners. The market leaders (frontrunners) are provided with insights, which allow them to keep ahead of the pack.

Key characteristics of frontrunners
The objective of the research was to identify the strategy style and organisational structures of innovation in a knowledge and service intensive sector. Those items, which make a difference in turning new ideas into profitable products and services. Remarkably frontrunners characterise themselves through their way of explicitly organising innovation, e.g. through a project team structure. This success factor enables the entrepreneurial role to become an innovation frontrunner. Also the use of external sources of knowledge to develop new products and services is a strength. The business cases of frontrunners have shown that customer involvement in innovation is far more complex than the literature review indicated.

What are the key success factors that make the frontrunners stay ahead?

First, of all the strategic focus and management priorities for new business development. Especially:

- the continuous attention for quality assurance in the service delivery;
- the concentration on core activities. This enables frontrunners to select potential profitable new product and service ideas and turn them into commercial viable services and products;
- early knowledge acquisitions of firms with new technologies or a unique market niche.

Secondly an innovation strategy style characterised by:

- being a trend-setter;
- investing more in R&D than competitors;
- the strong will to be first market entrant;
- the continuous search and investment for new business opportunities;
- anticipating environmental change.

Thirdly the resource capacity and allocation:

- the allocation of R&D resources to product innovation in stead of process innovation;
- the strong technological position and proficiency of the portfolio of technologies relative to competitors;
- knowledge development in cooperation with universities.

Fourthly, the organisation of innovation:

- by means of project team structure;
- with high quality predevelopment work and strong priority setting as part of the project planning of innovation projects;
- and associated cultural values: an internal competitive, non-hierarchical atmosphere and lack of conservatism.

Also, firms that address strategic issues at an early stage are more successful compared to firms timing certain issues at a later stage. With respect to timing, the images of this study are in line with findings in the literature. The following findings have been observed.

Successful innovators pay attention to more policy subjects at a time.
But we also observed that frontrunners draw attention to policy subjects and put them on the priority agenda at an earlier stage than less successful firms do. ICT firms that addressed the subject of cost reductions sooner on the management agenda are more likely to be a successful innovator. The same applies to the recruitment of personnel and training. The earlier ICT firms addressed these strategic issues, the better they score on the process success factor.

Frontrunners keep certain issues constantly at a high-priority level.
Success requires long-term commitment and devotion to key policy issues. The long-term and continuous devotion to product renewal is beneficial to ICT firms. Firms that practice this are successful in developing high-revenue and profit-generating products and services. Company B has institutionalised this with its clients both in the past and at present:

Software development is a continuous drive for company B, an international software firm. It has committed itself to launching new software releases in the market every two years. Alongside this goal, it aims to build proprietary software development tools. This requires a conscious choice of technologies, markets and customers. In 1985, it developed a third-generation software programming package one year ahead of its competitors. In 1986, a fourth-generation software tool accompanied the portfolio of software products. In 1989, an important choice was made to develop a fourth-generation software tool especially for ERP software (Enterprise Resource Planning). Company B's competitors were (still) working with traditional SDW methods (cascade model) and were using the existing development tools. Company B had invested a great deal of money in the new tool. At that moment company B had no resources left to invest in further product innovation, but the fourth generation ERP development tool was ready to be launched in the market and customers were waiting for it. The choice for open-system architecture is company B's differentiating factor. The choice for architecture enabled its clients to integrate the software with existing hardware, databases and interfaces.

The existing and potential clients of company B were aware that company B launched new software every two years and seemed satisfied with the new ERP tool. The timing of company B was even more remarkable for another reason than being ahead of competitors. The key feature of the 4GL ERP tool was its different organising principle for the development process. According to the marketing director, company B was ahead of its competitors in reorganising the internal software development structure: '... At a time where other companies were not thinking of doing the same, we decided to change the process to ensure rapid application development ...'. The company observed that the competitors started to reorganise two years later. Although company B also had to cope with decreasing sales figures for some of its existing products in the early 1990s, it rapidly grew both in size and at an international level afterwards. The value of the company increased rapidly on the stock market as stock prices grew by 100% every six months. The stock market seemed to appreciate company B's innovation strategy.

Successful innovators differentiate by addressing the following issues sooner.
The early emphasis on productivity growth, protection of intellectual assets and the acquisition of firms has distinguished the ICT frontrunners. Research (Cobbenhagen 1999, Cooper 1995) also considered this impact of timing differences in company policy, suggesting that a proactive policy contributes to success. The main thing is to do things ahead of competitors by concentrating on activities that are unique and hard to copy, similar to the arguments of dynamic capabilities theory.

Let us look at the nature and structure of the organisation. Apparently, integrating interdepartmental views through team-based structures was found to foster innovation success. Among the main organisation-related key success factors are:

- project team-based structures;
- cultural factors dominated by the absence of hierarchy and conservatism, and the presence of internal competitiveness;
- regular meetings between marketing and production and strong internal communication links between R&D, production and maintenance.

Such project matrix-like organisations can be more effective in enhancing project integration, speed and meeting client needs.

Above all, the nature of complex ICT projects requires a different type of organisation. Projects have short timeframes, tight schedules and complex technology integration. Companies are forced to complete projects on time and within the budget, particularly ICT infrastructure and software development projects. These firms typically encounter the difficulty of rapid application where the right technological choices and business choices must be made. At this crossroads of choices, a team approach helps to reduce the co-ordination problems. We learned from ICT firms that the organisation must be capable of structuring itself laterally at the project level, because:

- ICT firms are predominantly task-oriented.
- tasks have by definition a business and technology impact and are therefore multidisciplinary.
- tasks require specialist input in various stages of IT concept development, design and building and testing. This calls for involvement of several disciplines (designers, testers, maintenance) in various parallel stages of innovation.

The role of the customer: learning from customers

Customer focus is an important competitive characteristic for companies. However, to what extent is it a differentiating characteristic in actual practice? Qualitative data have indicated that customer focus is relevant to almost every ICT company. What then is of interest in customer interaction in this study? Many innovation studies fully support the idea that a company should be concentrating on its clients and involve them in new product development. Managers practise this outside-in thinking. In contrast with the mainstream innovation argument, we argue that too much customer involvement during the various phases of the innovation process could harm innovation success. This study observed that frontrunners use a different customer approach. Frontrunners play a different role in guiding customer involvement in innovation (Huizenga & Veldhoen, 2003). They do not focus on customer satisfaction, but deepen into the unaddressed and hidden needs of customers and try to provide solutions to customer problems. Customer involvement is especially relevant to frontrunners in two areas:

- improvement of quality;
- process enhancement.

Customer input is beneficial to improve service processes. Based on case study material the study revealed typical frontrunner approaches:

- frontrunners have a highly critical attitude towards finding unaddressed and hidden customer expectations and needs;
- frontrunners are aware of the danger inherent in serving conservative customers that feel no need to innovate;
- frontrunners build mutual trust in customer relationships but let customers pay for extra services.

This remarkable outcome seems to confirm the idea of restrictive co-operation and involvement with customers in various innovation stages in the ICT sector.

The selection environment
Results have also indicated that the service providers interpret the development of new services as the early absorption of new technological knowledge. The software developers place more emphasis on product development. Yet there seems to be no market condition that rewards service providers for investing in new technology.

These environmental circumstances have an impact on such issues as the quality of services and products, the role of R&D and scarce resources. The market circumstance might be favourable to almost all firms operating in the ICT sector. This suggests that there are only few firms that do not benefit from the growth of the sector. An actual selection mechanism (e.g. Hannah & Freeman) and market conditions that differentiate the high performers from the low performers might be absent (as illustrated in the case below for dispatching ICT personnel). Currently, ICT firms can grow in the short term due to a high ICT demand and short labour supply. There seems no need to commit to innovation. This study argued that quality was important to the ICT sector. This devotion to quality was observed in discussions with the managers and also revealed by the strategic priorities of ICT firms. Yet we found some exceptions of firms that were not fully committed to quality, but can still realise growth and profits in the short term. Due to the lack of market selection mechanisms, these firms are not competed out of the market.

Companies might incorporate firm characteristics, which the selection environment of today does not address or addresses only to a small extent. The market does not break into these characteristics, nor do these characteristics temporarily affect the financial results of a company. For example, a lack of investment in quality does not harm the revenues for the short term.

Company F posts ICT personnel at its customers. It has deliberately refrained from investing in training of its personnel for the last three years. The manager is explicit about the reasons for doing so: a direct loss of revenues is unacceptable if the employees are trained and do not make billable hours with the customer. This form of 'body shopping' results in short-term growth for company F. The management team of company F is aware of this fact and the downside of depreciating skills and knowledge. The company has a high employee turnover rate of 20%.

Next to company F, three other firms in the study have shown a similar practice. This might indicate that small outsourcing firms care less about the quality and skills of their employees but can still make good revenues in the short term. The market conditions do not rule out these practices.

8.2 RESEARCH AGENDA

This study is based on a dual research methodology: showcasing best practices and innovation issues from leaders and followers, using company surveys and case studies of more than 30 European firms. It included a cross sector perspective by contrasting the key success factors of ICT firms with findings from other studies. In this final section we speculate on the implications of the analysis for a broader and deeper understanding of the current role of innovation in the ICT sector. In this book we discussed a number of issues arising from the state of innovation affairs in the ICT sector. In order to contribute to the knowledge development on innovation management we essentially propose the following.

Innovation issues that need more attention
The contribution of the ICT innovation study is particularly in investigating and identifying a broad spectrum of sector-specific factors that explain innovation success. The dual research methodology is a worthwhile scientific area for business studies. The method has appeared to be transferable and applicable to several sectors. The study delivered the identification of a core set of key factors conducive to innovation success. But also the managerial insights 'why things happen'. We also succeeded to identify new success factors and confirm certain key success factors valid in other research studies, e.g. proficient pre-development work. The retrospective and replication nature of this study has strengthened the validity. What research areas need more attention?

First of all the attention of ICT innovation has merely focused on technological advancements, which were incorporated into novel hardware and software products. Service innovations were incremental and merely required to complement these technological innovations. More recently, the ICT sector has moved into an era in which knowledge and service-intensive organisations play a much more powerful role than previously. Particularly due to the need of grasping and exploring the customer needs. What can we conclude from this development? We observed that certain subjects are different from previous innovation studies, and that especially the observed complex role of customer interaction needs more research attention. The illustrative case material has shown that other factors play a role in customer interaction.

Whether we focus on the outputs of the service sectors or on the renewal of service functions, the growth of the service components is one of the great trends. A trend not only familiar to the ICT sector but to more industries and service sectors. More research is advocated on service innovations and the appearance of an innovation paradox ('do what the customer tells' versus 'tell the customer what to do'). Here the interactions with customers and suppliers are more complex than presumed in the literature. A second interesting research area. Thirdly, we learned that resource based arguments and path dependency could explain some of the contrary expectations. These items are among the major challenges for the innovation and strategy sciences. More insight is needed into the qualities and competencies that control and steer these innovation processes.

The innovation strategy style is another subject that deserves greater research attention. We have seen that many firms acknowledge the strategic relevance of quality and core competencies but at the same time that ICT firms differ in their timing of strategic priorities and organisation of innovation. Firms that concentrated their business activities on selecting profitable new product ideas, and turning innovation projects into commercial viable products and services were more successful. Research on the innovation strategy style might test assumptions as to whether frontrunners are more action-oriented, address priorities sooner and better foresee policy issues by picking up weak signals from the environment. Or investigate why frontrunners seem to have a more integral perspective of strategy and innovation. Paying more attention to all these ideas contributes to new knowledge development on innovation in the ICT sector.

Next research phase: towards configuration research
The outcomes of the ICT study convey the idea that the point of orchestration of the innovation process has shifted beyond multidisciplinary teams. But this movement is actually hard to address with a single key success factor. Next steps need to be taken. Single performance-oriented perspectives are no longer satisfactory to understand the 'why and how question of innovation'. Statistical constraints often force us to use success

notions to map innovation in a sector. However, we have to recognise that the understanding of innovation processes and consequences calls for combined performance- and process-oriented research, e.g. business-case research. No comparison can be made without performance research and no insight can be gained without understanding the business case. It is like blending words and numbers, e.g. a call for strategies that combine qualitative and quantitative methods for research (Hertog, Cobbenhagen, Huizenga, Bodewijs 2000). This will not be easy but the synergy in perspectives and methodologies has proved to be valuable in this study. The combination of business cases and management interviews complements the quantitative research. It also supports grasping developments around resource-based approaches to organisation, strategy and the role of the customer in innovation.

References

Abernathy, W.J. & Utterback, J., (1978), Patterns of Industrial Innovation, *Technology Review*, June-July, 40-47.

Abernathy, W.J. & Clark, K.B., (1985), Innovation: Mapping the Winds of Creative Destruction, *Research Policy*, **14**, 3-22.

Allen, T.J., (1978), *Managing the Flow of Technology*, MIT Press, Cambridge, MA.

Allen, T.J. & Cohen, S., (1969), Information Flow in R&D Laboratories, *Administrative Science Quarterly*, **14**, 19-24.

Allen, T.J. & Katz, R., (1986), The dual ladder: Motivational Solution or Managerial Delusion? *Research & Development Management*, **16** (2), 185-197.

Amburgey, T.L. & Dacin T., (1994), As the Left Foot Follows the Right? The Dynamics of Strategic and Structural Change, *Academy of Management Journal*, 37 (6), 1427-1452.

Amit, R. & Schoemaker, P.J.H., (1993), Strategic Assets and Organizational Rent, *Strategic Management Journal*, **14**, 33-46.

Ancona, D.G. & Caldwell D.F., (1992), Demography and Design: Predictors of New Product Team Performance, *Organization Science*, **3** (3), 321-341.

Andrews, K.R., (1971), *The Concept of Corporate Strategy*, Irwin, Homewood, IL.

Ansoff, I., (1965), *Corporate Strategy*, McGraw Hill, New York.

Argyris, C. & Schön, D.A., (1978), *Organisational Learning: A Theory of Action Perspective*, Addison-Wesley, Reading, MA.

Astley, W.G., (1984), Toward an Appreciation of Collective Strategy, *Academy of Management Review*, **9** (3), 526-535.

Barclay, I., (1992), The New Product Development Process: Part 1. Past Evidence and Future Practical Application, *R & D Management*, **22** (3), 255-263.

Barczak, G., (1995), New Product Strategy, Structure, Process, and Performance in the Telecommunications Industry, *Journal of Product Innovation Management*, **12**, 224-234.

Barney, J., (1986), Strategic factor markets: Expectation. Luck and Business Strategy, *Management Science*, **32** (10), 1231-1241.

Barney, J., (1991), Firm Resources and Sustained Competitive Advantage, *Journal of Management*, **17**, 99-120.

Bart, C.K., (1986), Product Strategy and Formal Structure, *Strategic Management Journal*, **7** (4), 293-312.

Becker, B. & Gerhard, B., (1996), The Impact of Human Resource Management on Organizational Performance. Progress and Prospects, *American Management Journal*, **39** (4), 779-802.

Biemans, W.G., (1992), *Managing Innovation within Networks*, Routledge, London.

Bolwijn, P.T. & Kumpe T., (1992) About Facts, Fiction and Forces in Human Resource Management, *Human Systems Management*, **15** (3), 161-172.

Bono, de E., (1994) *Serious Creativity: Using the Power of Lateral Thinking to Create New Ideas*, Harper Collins, London.

Booz, Allen & Hamilton, (1968), *Management of New Products*, Booz Allen & Hamilton Inc., New York.

Booz, Allen & Hamilton, (1997), *Enabling the Information Society: Supporting Market-Lead Developments*, Booz Allen & Hamilton, The Hague.

Bourgeois, L.J. III & Eisenhardt K. M., (1988), Strategic Decision Processes in High Velocity Environments: Four Cases in the Microcomputer Industry, *Management Science*, **34**, 816-835.

Bower, J.L., (1970), *Managing the Resource Allocation Process*, Boston, MA, Harvard University.

Brown, S.L. & Eisenhardt, K.M., (1995), Product Development: Past, Present and Future directions, *Academy of Management Review*, **20** (2), 343-378.

Brownell, D. & Dunk A. (1991), Task Uncertainty and its Interaction with Budgetary Participation and Budget Emphasis: Some Methodological Issues and Empirical Investigation, *Accounting, Organizations and Society*, **16** (8), 693.

Bruce, M. *et al.,* (1995), Success Factors for Collaborative Product Development: A study of Suppliers of Information and Communication Technology, *R&D Management*, **25** (1) 33-44.

Burgelman, R., (1983), A Model of the Interaction of Strategic Behaviour, Corporate Context, and the Concept of Strategy, *Academy of Management Review*, **1**, 61-70.

Burns, T. & Stalker, G.M., (1961), *The Management of Innovation*, Tavistock, London.

Calantone, R.J., Di Benedetto, C.A. & Divine, R., (1993), Organisational, Technical and Market Antecedents for Successful New Product Development, *R&D Management*, **23** (4), 337-351.

Cameron, K., (1980), Critical Questions in Assessing Organizational Effectiveness, *Organizational Dynamics*, **9**, 66-80.

Carter, C. & Williams B., (1957*), Industry and Technical Progress,* Oxford University Press.

Chaffee, E.E., (1985), Three Models of Strategy, *Academy of Management Review*, **10** (1), 89-98.

Chandler, A. D., (1962), *Strategy and Structure: Chapters in the History of the Industrial Enterprise*, Cambridge, MA, MIT Press.

Child, J., (1984), *Organisation: A Guide to Problems and Practice*, London: Harper & Row.

Christensen, C.R. *et al.,* (1978), Business Policy: Text and Cases, 6 ed. Homewoord, Irwin.

Clark, K. & Fujimoto, T., (1991), *Product Development Performance: Strategy, Organization and Management in the Auto Industry*, Harvard Business Press, Boston, MA.

Clark, K. & Wheelwright, S., (1993), *Managing New Product and Process Development*, The Free Press, New York.

Cobbenhagen, J., (1999), *Managing Innovation at the Company level: A Study of Non-sector Specific Success Sactors*, Datawyse - Universitaire Pers Maastricht.

Cobbenhagen, J., J. F. den Hertog & J.M. Pennings, (1995), *Succesvol Veranderen: Kerncompeteneties en Bedrijfsvernieuwing*, Kluwer BedrijfsWetenschappen, Deventer.

Cohen, W.M. & Levinthal, D.A., (1990), Absorptive Capacity: A New Perspective on Learning and Innovation, *Administrative Science Quarterly*, **35** (2), 128-152.

Collis, D.J., (1991), A Resource-Based Analysis of Global Competition: The Case of the Bearings Industry, *Strategic Management Journal*, **12**, 49-68.

Collis, D & Montgomery, C., (1995), Competing on Resources: Strategy in the 1999s, *Harvard Business Review*, **73** (4), 118-129.

Coombs, R., (1999), *Innovations in Services*, paper presented at Nijmegen Business School, 1999. Manchester: Manchester School of Management UMIST/CIRC.

Cooper, R., (1979), Identifying Industrial New Product Success: Project NewProd, *Industrial Marketing Management*, **8**, May.

Cooper, R.G., (1994), Third Generation New Product Processes, *Journal of Product Innovation Management*, **11** (1), 3-15.

Cooper, R.G., (1996), *Winning at New Products: Accelerating the Process from Idea to Launch*, Addison-Wesley, reading, MA.

Cooper, R. & Kleinschmidt, E.J., (1987), What Makes a New Product a Winner: Success Factors at the Project Level, *R&D Management*, **17** (3), 175-189.

Cooper, R. & Kleinschmidt, E.J., (1993), Screening New Products for Potential Winners, *Long Range Planning*, **26**, (6), 74.

Cooper, R. & Kleinschmidt, E.J., (1993), New-Product Success in the Chemical Industry, *Industrial marketing management*, **22** (2), 85.

Cooper, R.G. & Kleinschmidt, E.J., (1995), Benchmarking Firm's Critical Success Factors in New Product Development, *Journal of New Product Development*, **12** (5), 374-391.

Cooper, R.G. & Kleinschmidt, E.J., (1996), Winning Businesses in Product Development: The Critical Success Factors, *Research-Technology Management* , **39** (4), 18-29.

Cooper, R.G. & Moore, R.A., (1979), Modular Risk Management: an Applied Example, *R&D Management*, 9 (2), 93-99.

Cusumano, M. & Selby, R., (1995), *Microsoft Secrets: How the World's Most Powerful Software Company Creates Technology, Shapes Markets and Manages People*, HarperCollins Business, London.

Daft, R.L., (1983), Learning the Craft of Organizational Research, *Academy of Management Review*, **8** (4), 539-546.

Daft, R. & Lewin, A., (1993), Where are the Theories of the 'New' Organisation Forms? An Editorial Essay, *Organisation Science*, **4**, (4) i-iv.

Damanpour, F., (1991), Organizational Innovation: A Meta-analysis of Effects of Determinants and Moderators, *Academy of Management Journal*, **34** (3) 555.

D'Aveni, R., (1996), Hyper-competition: Managing the Dynamics of Strategic Maneuvering, *Academy of Management Review*, **21** (1), 291-293.

Delery, J. & Doty, D., (1996), Modes of Theorizing in Strategic Human Resource Management: Tests of Universalistic, Contingency, and Configurational Performance Predictions, *Academy of Management Journal*, **39** (4), 802-835.

Dierckx, I. & Cool, K., (1989), Asset Stock Accumulation and Competitive Advantage, *Management Science*, **12**, 1504-1511.

Donaldson, L., (1995), *American Anti-management Theories of Organization: A Critique of Paradigm Proliferation*, Cambridge University Press, Cambridge.

Doty, D. & Glick, A. & Huber G., (1993), Fit, Equifinality, and Organizational Effectiveness: A Test of Two Configurational Theories, *Academy of Management Journal*, **36**, (6), 1196.

Dougherty, D., (1990), Understanding New Markets for New Products, *Strategic Management Journal*, **11**, 179-202.

Dougherty, D., (1992a), A Practice-Centered Model of Organizational Renewal through Product Innovation, *Strategic Management Journal*, **13**, 77.

Dougherty, D., (1992b), Interpretive Barriers to Successful Product Innovation in Large Firms, *Organization Science*, **3**, 179-202.

Dougherty, D. & Corse, S.M., (1995), When It Comes to Product Innovation, What Is So 'Bad' About 'Bureaucracy'?, *Journal of High Technology Management.*

Dougherty, D. & Hardy, C., (1996), Sustained Product Innovation in Large, Mature Organizations: Overcoming Innovation-to-Organization Problems, *Academy of Management Journal*, **39** (5), 1120-1153.

Downs, G. & Mohr, L., (1976), Conceptual Issues in the Study of Innovation, *Administrative Science quarterly*, **21**, 700-714.

Drucker, P., (1993), *Post Capitalist Society*, Harper Business, New York.

Dwyer, L. *et al.,* (1991), Organizational Environment, New Product Process Activities, and Project Outcomes, *Journal of Product Innovation Management*, **8**, 39-48.

Dyer, A. & B. Wilkins, (1991), Better Stories, Not Better Bonstructs to Generate Better Theory: A Rejoinder to Eisenhardt, *Academy of Management Review*, **16** (3), 613-619.

Eisenhardt, K.M., (1989), Building Theories From Case Study Research, *Academy of Management Review*, **14** (4), 532-550.

Eisenhardt, K. & B. Tabrizi (1995), Accelerating Adaptive Processes: Product Innovation in the Global Computer Industry, *Administrative Science Quarterly*, **40** (1) 84-110.

EITO, (1999), *European Information Technology Observatory*, Frankfurt am Main.

Fayol, H., (1949), General Principles of Management, in: Shadritz & Ott (eds.), (1987): *Classics of Organization Theory*, Dorsey Press.

Fowler, F.J., (1988), *Survey Research Methods* (revised edition), Sage, Newbury Park.

Freeman, C., (1988), *The Economics of Industrial Innovation*, 2nd edition, London: Frances Pinter.

Galbraith, J.R., (1973) *Designing Complex Organizations*, Addison Wesley, Reading, MA.

Galbraith, J.R., (1994), *Competing with Flexible Lateral Organizations*, 2nd edition, Addison Wesley, Reading, MA.

Galbraith, J.R. & Nathanson, D.A., (1979), *Strategy Implementation: The Role of Structure and Process*, West, St. Paul MN.

Gersick, C., (1994), Pacing Strategic Change: The Case of a New Venture, *Academy of Management Journal*, **37**, (1), 9.

Giddens, A., (1979), *Central Problems in Social Theory*, Macmillan, London.

Glaser, B. & Strauss, A., (1967), *The Discovery of Grounded Theory: Strategies of Qualitative Research*, Wiedenfeld & Nicholson, London.

Gobeli, D. & Brown, D., (1988), Analyzing Product Innovations, *Research Management*, July-August, 25-31.

Gomez-Mejia, L. & Balkin, D., (1989), Effectiveness Of Individual and Aggregate Compensation Strategies, *Industrial Relations*, **28**, 431-445.

Grant, R.M., (1991), The Resource-Based Theory of Competitive Advantage: Implications for Strategy Formulation, *California Management Review*, 114-135, Spring.

Grant, R.M., (1996), Prospering in Dynamically-competitive Environments: Organizational Capability as Knowledge Integration, *Organization Science*, **7**, (4), 375-387.

Griffin, A. & Hauser, J., (1996), Integrating R&D and marketing : a review and analysis of the literature, *Journal* of *Product Innovation Management*, **13**, (3) 191.

Griffin, A. & Page, L. (1996), PDMA Success Measurement Project: Recommended Measures for Product Development Success and Failure, *Journal* of *Product Innovation Management*, **13** (6), 478.

Hall, D. & Saias, M., (1980), Strategy Follows Structure, *Strategic Management Journal*, **1**, 149-163.

Hamel, G. & Heene, A., (1994), *Competence-based Competition.*, John-Willey, Chichester.

Hamel, G. & Prahalad, C.K., (1991), Strategic Intent, *Harvard Business Review*, May-June, 63-76.

Hamel, G. & Prahalad, C.K., (1994), *Competing for the Future*, Harvard Business School Press, Boston, MA.

Hannah, M.T. & Freeman, J.H., (1989), *Organizational Ecology*, Harvard University Press, Cambridge.

Hansen, G. & Wernerfelt, B., (1989), Determinants of Firm Performance: The Relative Importance of Economic and Organisational Factor, *Strategic Management Journal*, **10**, 399-411.

Hedlund, G. & Rolander, D., (1987), *The Strategy-Structure Paradigm in International Business Research and Practice*. Research paper 87/4, Institute of International Business at the Stockholm School of Economics, Stockholm.

Heene, A. & Sanchez, R., (1997), *Competence-based Strategic Management*, John Wiley & Sons, Chichester.

Henderson, R. & Clark, K., (1990), Architectural Innovation: The Reconfiguration of Existing Product Technologies and the Failure of Established Firms, *Administrative Science Quarterly*, **35**, 9-30.

Henderson, R. & Cockburn, I., (1996), Scale, Scope, and Spillovers: The Determinants of Research Productivity in Drug Discovery, *The Rand Journal of Economics*, **27** (1), 32-59.

Hertog, J.F. Den & van Sluijs E., (1995a), Managing Knowledge Flows: A Key Role For Personnel Management, in: Europe's Next Step: *Organizational Innovation, Competition and Employment*, Eds. Andreasen, L.E., Coriat. B., den Hertog, J.F. and Kaplinsky, R. Frank Cass, Ilford.

Hertog, J.F. Den & van Sluijs E., (1995b), *Research in Organisations. A Methodological Travel Guide*, Van Gorcum, Assen (in Dutch).

Hertog, J.F. Den & Huizenga, E.I., (1997a), *The Innovative Software Enterprise: Strategy, Organisation and Personnel Policy*, Kluwer Bedrijfswetenschappen, Deventer (in Dutch).

Hertog, J.F. Den & Huizenga, E.I., (1997b), *The Knowledge Factor: Competing as Knowledge Enterprise*, Kluwer Bedrijfsinformatie, Deventer (in Dutch).

Hertog, J.F. Den & Huizenga, E.I., (2000), *The Knowledge Enterprise: Implementation of Intelligent Business Strategies*, Imperial College Press, London.

Hertog, J.F., Den & Cobbenhagen, J., Huizenga, E., Bodewijs, W., (2000), *Blending Words and Numbers: Strategies For Combining Qualitative and Quantitative Methods For Organizational Research*, MERIT, Maastricht University.

Hitt, M., Ireland. R., (1985), Corporate Distinctive Competence, Strategy, Industry and Performance, *Strategic Management Journal*, **6**, 273-293.

Hoopes, D.G. *et al.*, (2003) Why Is There a Resource-Based View? Toward a Theory of Competitive Heterogeneity, *Strategic Management Journal*, special issue, **24** (10), 889-1068.

Huizenga, E.I. (1997), *The Knowledge Factor: Competitive Advantage Through Knowledge Management*, conference paper presented at Annual Conference Strategic Management Society, IESE Business School, Barcelona.

Huizenga, E.I. (2001), *Innovation Management: How Frontrunners Stay Ahead, An Empirical Study on Key Success Factors in the ICT Sector*, Datawyse - Universitaire Pers Maastricht.

Huizenga, E.I. (2002a), *Industry Restructuring? Strategic Timing and Path Dependency Differentiate the Innovation Leaders,* conference paper presented at Strategic Management Society, Annual Conference, Paris.

Huizenga, E.I. (2002b), *New Business Development: Organisational Performance and Knowledge Strategies in The ICT-sector,* conference paper for Academy of Management Annual Meeting, Business Policy and Strategy Division.

Huizenga, E.I. & Veldhoen, B., (2003) Listening to Customers: Basis for Renewal (in Dutch), *Marketing Tribune,* **20** (22), 12-13.

Huselid, M. *et al.,* (1997), Technical and Strategic Human Resource Management Effectiveness as Determinants of Firm Performance, *Academy of Management Journal,* **40** (1), 171-188.

Huselid, M. & Becker, B., (1996), Methodological Issues in Cross-Sectional and Panel Estimates of the Human Resource-Firm Performance Link, *Industrial Relations,* **35** (3), 400-422.

Iansiti, M. & Clark, K., (1994), Integration and Dynamic Capability: Evidence from Product Development in Automobiles and Mainframe Computers, *Industrial and Corporate Change,* **3** (3),557-606.

Iansiti, M. & Khanna, T., (1995), Technological Evolution, System Architecture and The Obsolescence of Firm Capabilities, *Industrial and Corporate Change,* **4** (2), 333-362.

Imai, K., Nonaka, I. & Takeuchi, H., (1985), Managing The New Product Development Process: How Japanese Companies Learn and Unlearn, in: K. Clark *et al.* (eds.), *The Uneasy Alliance*, Harvard Business School Press, Boston.

Jensen, M. & Meckling, W., (1976), Theory of the Firm: Managerial Behavior, Agency Costs and Ownership Structure, in: Werin, L. & H. Wijkander (eds.), Contract Economic, Blackwell, Oxford, 251-274.

Johne, A. & Snelson, P., (1988), Success Factors In Product Innovation: A Selected Review Of The Literature, *Journal of Product Innovation Management,* **5** (2), 100-110.

Karlsson, C. & Ahlstrom, A., (1996), The Difficult Path To Lean Product Development, *Journal of Product Innovation Management,* **13,** (4).

Keller, R.T., (1986), Predictors Of The Performance Of Project Groups In R&D Organisations, *Academy of Management Journal,* **29** (4), 715-726.

Ketchen, D. *et al.,* (1997), Organizational Configurations and Performance: A Meta-Analysis, *Academy of Management Journal,* **40** (1), 222-240.

Kets de Vries, M., (1991), Whatever Happened To The Philosopher-King? The Leader's Addiction To Power, *Journal of Management Studies*, **28** (4).

Khandwalla, P., (1973), *The Design of Organizations*, Harcourt, Brace, Jovanovich, New York.

Klimstra, P.D. & Pots, J., (1988), What We've Learned About Managing R&D Projects, *Research & Technology Management*, May-June, 42-58.

Kline, S.J., (1985), Innovation Is Not A Linear Process, *Research Management*, July-August, 36-45.

Kogut, B. and Zander, U., (1995). Knowledge Of The Firm, Combinative Capabilities, And The Replication Of Technology, *Organization Science,* **3** (3), 383-397.

Larson, E.W. & Gobeli, D.H., (1985), Project Management Structures: Is There A Common Language? *Project Management Journal*, **16**, (2), 40-44.

Larson, E.W. & Gobeli, D.H., (1988), Organizing for Product Development Projects, *Journal of Product Innovation Management*, **5**, 180-190.

Larsson, R., (1993), Case Survey Methodology: Qualitative Analysis Of Patterns Across Case Studies, *Academy of Management Journal*, **36** (6), 1515-1546.

Lawrence, P.R. & Lorsch, J.W, (1967), *Organization and Environment: Managing Differentiation and Integration*, Irwin, Homewood.

Leonard-Barton, D., (1990), A Dual Methodology for Case Studies: Synergist Use of a Longitudinal Single Site With Replicated Multiple Sites, *Organization Science*, **1** (3), 248-66.

Leonard-Barton, D., (1995), *Wellsprings of Knowledge*, Harvard Business School Press, New York.

MacDuffie, J., (1995), Human Resource Bundles and Manufacturing Performance: Organizational Logic and Flexible Production Systems In The World Auto Industry, *Industrial & Labor Relations Review*, **48** (2), 197-221.

Mahoney, J. & Pandian, J., (1992), The Resource-based View Within the Conversation of Strategic Management, *Strategic Management Journal*, **13**, (5), 363.

Maidique, M.A. & Zirger, B.J., (1985), The New Product Learning Cycle, *Research Policy*, **14**, 299-313.

McCann & Galbraith, J. in: Cobbenhagen, J., (1999), *Managing Innovation At The Company Level: A Study Of Non-sector Specific Success Factors*, Datawyse/Universitaire Pers Maastricht, Maastricht.

McGrawth, R.G., (1995), Creating and Transforming Core Business: A Competence Oriented Analysis, paper Academy of Management Annual Meeting 1995.

McGrawth, R.G., (1997), A Real Options Logic for Initiating Technology Positioning Investments, *The Academy of Management Review*, **22** (4), 974-996.

McGrawth, R.G., MacMillan, I., & Venkatraman, S., (1995), Defining and Developing Competence: A Strategic Process Paradigm, *Strategic Management Journal*, **16**, 251-275.

Miles, R.E. & Snow, C.C., (1978), *Organizational Strategy, Structure and Process*, McGraw-Hill, New York.

Miller, D., (1986) Configurations of Strategy and Structure: Towards a Synthesis, *Strategic Management Journal*, 7, 233-249.

Miller, D. & Friesen, P., (1980), Archetypes of Organisational Transition, *Administrative Science Quarterly*, **15** (2), 268-300.

Mintzberg, H., (1979a), An Emerging Theory For Direct Research, *Administrative Science Quarterly*, **24**, 582-589.

Mintzberg, H., (1979b), *The Structuring of Organisations*, Prentice Hall, Englewood Cliffs, New York.

Mintzberg, H., (1994), Strategy Formation: Schools of Thought, in: Frederickson, J.W., *Perspectives on Strategic Management*, New York, 105-235.

Miyazaki, H. *et al.*, (1995) On the Level Probabilities for Useful Partially Ordered Alternatives in the Analysis of Variance, *Communications in Statistics*, **24** (8), 2059-2072.

Moeneart, R. *et al.*, (1990), R&D Marketing Integration Mechanisms, Communication Flows and Innovation Success, *Journal of Product Innovation Management*, **11**, 31-45.

Mohr, L.B., (1982), *Explaining Organizational Behaviour: The Limits and Possibilities of Theory and Research*, San Francisco: Jossey-Bass.

Mohrman, S. *et al.*, (1992), Human Resource Strategies For Lateral Integration In High Technology Settings, in: *Human Resource Strategy in High Technology*, Gomez-Mejia, L.R. & Lawless, M.W., JAI Press, Greenwich, CT.

Montoya-Weiss, M.M. & Calantone, R., (1994), Determinants Of New Product Performance: A Review and Meta-analysis, *Journal of Product Innovation Management*, **11**, 397-417.

Moss Kanter, R.M., (1983), *The Change Masters: Innovation for Productivity in the American Corporation*, Simon and Schuster, New York.

Myers, S. & Marquis, D., (1969), *Successful Industrial Innovation*, Washington D.C., National Science Foundation.

Nambisan, S. (2001), Why Services Businesses Are Not Product Businesses, *MIT Sloan Management* Review, **42**, (2), 72-80.

Nelson, R. & Winter, S., (1982), *An Evolutionary Theory of Economic Change*, Cambridge, Harvard University Press.

Nonaka, I., (1990), Redundant, Overlapping Organization: A Japanese Approach to Managing the Innovation Process, *California Management Review*, **32** (3), 27-38.

Nonaka, I., (1991), A Dynamic Theory Of Organizational Knowledge Creation, *Organization Science*, **5** (1), 14-37.

Nonaka, I. & Takeuchi, H., (1996), *The Knowledge Creating Company*, Oxford University Press.

OECD, (1994), *Frascati Manual: The Measurement Of Scientific And Technological Activities: Proposed Standard Practice For Surveys Of Research And Experimental Development*, Paris.

O'Reilly, C. *et al.*, (1991), People and Organizational Culture: A Profile Comparison Approach to Assessing Person-Organization Fit, *Academy of Management Journal*, **34** (3), 487.

Pavitt, K., (1991), Key Characteristics of the Large Innovating Firm, *British Journal of Management*, **2**, 41-50.

Pennings, J.M. & F. Harianto, (1992), The Diffusion of Technological Innovation in the Commercial Banking Industry, *Strategic Management Journal*, **13** (1), 29.

Penrose, E., (1959), *The Theory Of The Growth Of The Firm*, Oxford: Basil Blackwell.

Peteraf, M., (1993), The Cornerstones of Competitive Advantage: A Resource-based View, *Strategic Management Journal*,.**14** (3), 179

Peters, T., (1988), The Mythology Of Innovation, Or A Skunkwork Tale, Part II, in: Tushman, M.L., & Moore, W.L., *Readings in The Management of Innovation*, Harper Business, 138-148.

Pettigrew, A., (1985), *The Management of Strategic Change*, Oxford, Basil Blackwell.

Pettigrew, A.M., (1990), Longitudinal Field Research On Change: Theory And Practice, *Organization Science*, **1**, (3), 267-292.

Pfeffer, J., (1981), *Power In Organisations*, Marshfield, MA, Pitman.

Pfeffer, J., (1994), *Competitive Advantage through People*, Harvard Business Press, Boston.

Pierce, J.L. & Delbecq, A.L., (1977), Organization Structure, Individual Attitudes, And Innovation, *Academy of Management Review*, 27-37.

Pisano, G.P. & S.C. Wheelwright, (1995), The Logic of High-tech R&D, *Harvard Business Review*, **73** (5), 93-107.

Polanyi, M., (1966), *The Tacit Dimension*, Anchor Books, New York.

Porter, M., (1980), *Competitive Strategy: Techniques for Analyzing Industries and Competitors*, The Free Press, New York.

Porter, M., (1985), *Competitive Advantage: Creating and Sustaining Superior Performance*, The Free Press, New York.

Prahalad, C.K. & Hamel, G., (1990), The Core Competence Of The Organisation, *Harvard Business Review*, **68** (3), 79-91.

Prahalad, C.K. & Hamel, G., (1994), Strategy As A Field Of Study: Why Search For A New Paradigm? *Strategic Management Journal*, **15**, 5-16.

Pugh, D., Hickson, D., Hinnings, C. & Turner, C., (1969), Dimensions Of Organisation Structure, *Administrative Science Quarterly*, **13**, 65-105.

Quinn, J.B., (1985), Managing Innovation: Controlled Chaos, *Harvard Business Review*, May-June, 73-84.

Quinn, J.B., (1993), Managing The Intelligent Enterprise: Knowledge & Service-Based Strategies, *Planning Review*, **21** (5), 123-137.

Rothwell, R., (1972), *Factors for Success in Industrial Innovation. Project SAPPHO: A Comparative Study Of Success And Failure In Industrial Innovation*, Brighton, Sussex , SPRU.

Rothwell, R., (1974), SAPPHO- Updated – Project SAPPHO Phase II, *Research Policy*, (3), 30-38.

Rothwell, R., (1992), Successful Industrial Innovation: Critical Factors in the 1990s, *R&D Management*, **22** (3).

Rothwell, R., (1994), Industrial Innovation, Success, Strategy, Trends, in: Dodgeson, M., & Rothwell, R., *The Handbook of Innovation Management*, Edward Elgar, Aldershot.

Roussel, Ph. A. *et al.*, (1991), *The Third R&D Generation, Managing The Link To Corporate Strategy*, Harvard Business School Press, Boston.

Rumelt, R.P, (1984), Towards A Strategic Theory of the Firm, in: Lamb, R.B. ed., *Competitive Strategic Management*, Prentice Hall, Englewood-Cliffs, NJ.

Rumelt, R.P, (1991), How Much Does Industry Matter?, *Strategic Management Journal*, **12** (3), 167.

Rumelt, R. & Schendel, D., Teece, D., (1996), Fundamental Issues in Strategy: A Research Agenda, *Administrative Science Quarterly*, **41**, (1), 196-198.

Sanchez R. & Heene, A., (1997), Reinventing Strategic Management: New Theory and Practice for Competence-based Competition, *European Management Journal*, **15**, (3), 303-317.

Sanchez, R. & Heene, A., Thomas, H., (1996), *Dynamics of Competence-Based Competition: Theory And Practice In The New Strategic Management*, Pergamon, Oxford.

Sarren, M., (1984), A Classification And Review Of Models Of The Intra-Innovation Process, *R&D Management*, **14** (1), 11-24.

Schein, E.H., (1985), *Organizational Culture and Leadership,* Jossey-Bass, San Francisco.

Schmalensee, R., (1985), Do Markets Differ Much?, *The American Economic Review*, **75**, 341-350.

Schreuder, H. *et al.,* (1991), Successful Bear-Fighting Strategies, *Strategic Management Journal*, **12**, (7), 523.

Schuler, R.S. & Jackson, S.E., (1996), *Human Resource Management, Positioning for the 21st century* (6th ed.), West Publishing Company, Minneapolis, MN.

Schumpeter, J.A., (1942), *Capitalism, Socialism And Democracy*, Harper and Row, New York.

Scott-Morgan, P., (1994), *The Unwritten Rules of the Game*, McGraw-Hill, New York.

Senge, P.M., (1990), *The Fifth Discipline: The Art and Practice Of The Learning Organization*, Century Business, London.

Shortell, S.M. & Zajac, E.J., (1990), Perceptual and Archival Measurement of Miles and Snow Strategic Types: A Comprehensive Assessment Of Reliability And Validity, *Academy of Management Journal*, **33** (4), 817-832.

Shrivastava, P. & Souder, W.E., (1987), The Strategic Management Of Technological Innovations: A Review And Model, *Journal of Management Studies*, **24** (1), January.

Snow, C. & Hrebiniak, L., (1980), Strategy, Distinctive Competence, And Organizational Performance, *Administrative Science Quarterly*, **25**, 317-335.

Song, X. & Parry, M., (1996), What Separates Japanese New Product Winners From Losers, *Journal of Product Innovation Management*, **13** (5), 422.

Song, X. & Parry, M., (1999), Challenges Of Managing The Development Of Breakthrough Products In Japan, *Journal of Operations Management*, **17** (6), 665-688.

Souder, W.E., (1987), *Managing New Product Innovations*, Lexington Books, New York.

Stacey, R., (1996), *Strategic Management & Organisational Dynamics*, 2nd edition, Pitman, London.

Stopford, J. & Baden-Fuller, C., (1994), Creating Corporate Entrepreneurship, *Strategic Management Journal*, **15** (7), 521-536.

Strauss, A. & Corbin, J., (1990), *Basics of Qualitative Research: Grounded Theory, Procedures and Techniques*, Sage, Newbury Park.

Szulanski, G., (1996), Exploring Internal Stickiness: Impediments to the Transfer of Best Practice Within the Firm, *Strategic Management Journal*, **17**, 27-44.

Takeuchi, H. & Nonaka, I., (1986), The New Product Development Game: Stop Running The Relay Race And Take Up Rugby, *Harvard Business Review*, **64** (1), 137-146.

Taylor, F.W., (1911), The Principles of Scientific Management, in: Shadritz & Ott (eds.), (1987): *Classics of Organization Theory*, Dorsey Press.

Teece, D. & Pisano, G., (1994), The Dynamic Capabilities of Firms: An Introduction, *Industrial and Corporate change*, **3** (3), 537-556.

Teece, D., Pisano, G. & Shuen, A., (1997), Dynamic Capabilities and Strategic Management, *Strategic Management Journal*, **18** (7), 509-533.

Thomke, B.S. & Von Hippel, E., (2002), Customers As Innovators: A New Way To Create Value, *Harvard Business Review*, **80**, 74-81.

Tushman, M. & Anderson, P., (1986), Technological Discontinuities and Organisational Environments, *Administrative Science Quarterly*, **31**, 439-465.

Tushman, M. & Moore, W.L., (1992), *Readings In The Management Of Innovation*, Harper Business, New York.

Twiss, B., (1980), *Managing Technological Innovation*, 4th edition, London, Pitman.

Vasconcellos, J. & Hambrick, D., (1989), Key Success Factors: Test Of A General Framework In The Mature Industrial Product Sector, *Strategic Management Journal*, **10**, 367-382.

Vaus, D.A. De (1991), *Surveys in Social Research*, 3rd Edition, UCL Press/Allen & Uniwin, London.

Ven, A. van de, (1986), Central Problems In The Management Of Innovation, *Management Science*, **32** (5), 590-607.

Ven, A. van de & Poole, S., (1989), *Research On The Management Of Innovation: The Minnesota Studies*, Harper Row, New York.

Ven, A. van de & G. Huber, (1990), Longitudinal Field Research Methods For Studying Processes Of Organizational Change, *Organization Science,* **1** (3), 213-219.

Von Hippel, E., (1978), A Customer Active Paradigm for Industrial Product Idea Generation, *Research Policy*, 240-266.

Von Hippel, E., (1986), Lead Users: A Source of Novel Product Concepts, *Management Science*, **32**, 7.

Voss, A.C., (1988), Implementation: A Key Issue In Manufacturing Technology: The Need For A Field for Study, *Research Policy*, **17**, 55-63.

Weber, M., (1949), *The Theory Of Social And Economic Organization*, in: Shadritz, R. & Ott, A., (eds.), (1987): *Classics of Organization Theory*, Dorsey Press.

Wernerfelt, B., (1984), A Resource-Based View of the Firm, *Strategic Management Journal*, **5** (2), 171-180.

Whittington, R. *et al.,* (1993), What Is Strategy -- And Does It Matter?, *Organization Studies*, **15** (6), 931-933.

Williamson, O., (1975), *Markets & Hierarchies*, Free Press, New York.

Womack, J., Jones, D., & Roos, D., (1990), *The Machine That Changed The World*, Rawson Associates, Maxwell Macmillan, New York.

Woodward, J., (1965), *Industrial Organization: Theory And Practice*, Oxford University Press, London.

Wright, P., McMahan, G., McCormick, B., Sherman, W., (1998), Strategy, Core Competence, And HR Involvement As Determinants Of HR Effectiveness And Refinery Performance, *Human Resource Management*, **37** (1), 17-30.

Yin, R.K., (1989), *Case Study Research, Design And Methods*, Sage, Beverly Hills.

Zmud, R.W., (1982), Diffusion of Modern Software Practices: Influence Of Centralization And Formalization, *Management Science*, **28**, 1421-1431.

Index

Author index